CAMBRIDGE TRACTS IN MATHEMATICS

General Editors

B. BOLLOBAS, W. FULTON, A. KATOK, F. KIRWAN,
P. SARNAK, B. SIMON, B. TOTARO

166 The Lévy Laplacian

The Lévy Laplacian is an infinite-dimensional generalization of the well-known classical Laplacian. Its theory has been increasingly well-developed in recent years and this book is the first systematic treatment of it.

The book describes the infinite-dimensional analogues of finite-dimensional results, and more especially those features that appear only in the generalized context. It develops a theory of operators generated by the Lévy Laplacian and the symmetrized Lévy Laplacian, as well as a theory of linear and nonlinear equations involving it. There are many problems leading to equations with Lévy Laplacians and to Lévy–Laplace operators, for example superconductivity theory, the theory of control systems, the Gauss random field theory, and the Yang–Mills equation.

The book is complemented by exhaustive bibliographic notes and references. The result is a work that will be valued by those working in functional analysis, partial differential equations and probability theory.

Cambridge Tracts in Mathematics

All the titles listed below can be obtained from good booksellers or from Cambridge University Press. For a complete series listing visit http://publishing.cambridge.org/stm/mathematics/ctm/

The Lévy Laplacian

M. N. FELLER

CAMBRIDGE
UNIVERSITY PRESS

CAMBRIDGE UNIVERSITY PRESS
Cambridge, New York, Melbourne, Madrid, Cape Town, Singapore, São Paulo

Cambridge University Press
The Edinburgh Building, Cambridge CB2 2RU, UK

Published in the United States of America by Cambridge University Press, New York

www.cambridge.org
Information on this title: www.cambridge.org/9780521846226

First published 2005

Printed in the United Kingdom at the University Press, Cambridge

A record for this publication is available from the British Library

Library of Congress in Publication data

ISBN-13   978-0-521-84622-6 hardback
ISBN-10   0-521-84622-6 hardback

Contents

Introduction

The Laplacian acting on functions of finitely many variables appeared in the works of Pierre Laplace (1749–1827) in 1782. After nearly a century and a half, the infinite-dimensional Laplacian was defined. In 1922 Paul Lévy (1886–1971) introduced the Laplacian for functions defined on infinite-dimensional spaces.

The infinite-dimensional analysis inspired by the book of Lévy *Leçons d'analyse fonctionnelle* [93] attracted the attention of many mathematicians. This attention was stimulated by the very interesting properties of the Lévy Laplacian (which often do not have finite-dimensional analogues) and its various applications.

In a work [68] (published posthumously in 1919) Gâteaux gave the definition of the mean value of the functional over a Hilbert sphere, obtained the formula for computation of the mean value for the integral functionals and formulated and solved (without explicit definition of the Laplacian) the Dirichlet problem for a sphere in a Hilbert space of functions. In this work he called harmonic those functionals which coincide with their mean values.

In a note written in 1919 [92], which complements the work of Gâteaux, Lévy gave the explicit definition of the Laplacian and described some of its characteristic properties for the functions defined on a Hilbert function space.

In 1922, in his book [93] and in another publication [94] Lévy gave the definition of the Laplacian for functions defined on infinite-dimensional spaces and described its specific features. Moreover he developed the theory of mean values and using the mean value over the Hilbert sphere, solved the Dirichlet problem for Laplace and Poisson equations for domains in a space of sequences and in a space of functions, obtained the general solution of a quasilinear equation. We have mentioned here only a few of a great number of results given in Lévy's book which is the classical work on infinite-dimensional analysis.

The second half of the twentieth century and the beginning of twenty-first century follows a period of development of a number of trends originated in [93], and the infinite-dimensional Laplacian has become an object of systematic study. This was promoted by the appearance of its second edition *Problèmes concrets d'analyse fonctionnelle* [95] in 1951 and the appearance, largely due to the initiative of Polishchuk, of its Russian translation (edited by Shilov) in 1967. During this period, there were published, among others, the works of: Lévy [96], Polishchuk [111–125], Feller [36–66], Shilov [132–135], Nemirovsky and Shilov [102], Nemirovsky [100, 101], Dorfman [28–33], Sikiryavyi [137–145], Averbukh, Smolyanov and Fomin [10], Kalinin [82], Sokolovsky [146–151], Bogdansky [13–22], Bogdansky and Dalecky [23], Naroditsky [99], Hida [75–78], Hida and Saito [79], Hida, Kuo, Potthoff and Streit [80], Yadrenko [158], Hasegawa [72–74], Kubo and Takenaka [85], Gromov and Milman [69], Milman [97, 98], Kuo [86–88], Kuo, Obata and Saito [89, 90], Saito [126–129], Saito and Tsoi [130], Obata [103–106], Accardi, Gibilisco and Volovich [4], Accardi, Roselli and Smolyanov [5], Accardi and Smolyanov [6], Accardi and Bogachev [1–3], Zhang [159], Koshkin [83, 84], Scarlatti [131], Arnaudon, Belopolskaya and Paycha [9], Chung, Ji and Saito [26], Léandre and Volovich [91], Albeverio, Belopolskaya and Feller [8].

Many problems of modern science lead to equations with Lévy Laplacians and Lévy–Laplace type operators. They appear, for example, in superconductivity theory [24, 71, 152, 155], the theory of control systems [121, 122], Gauss random field theory [158] and the theory of gauge fields (the Yang–Mills equation) [4], [91].

Lévy introduced the infinite-dimensional Laplacian acting on a function $U(x)$ by the formula

$$\Delta_L U(x_0) = 2 \lim_{\varrho \to 0} \frac{\mathfrak{M}_{(x_0, \varrho)} U(x) - U(x_0)}{\varrho^2}$$

(the Lévy Laplacian), where $\mathfrak{M}_{(x_0, \varrho)} U(x)$ is the mean value of the function $U(x)$ over the Hilbert sphere of radius ϱ with centre at the point x_0.

Given a function defined on the space of a countable number of variables we have

$$\Delta_L U(x_1, \ldots, x_n, \ldots) = \lim_{n \to \infty} \frac{1}{n} \sum_{k=1}^{n} \frac{\partial^2 U}{\partial x_k^2},$$

while for functions defined on a functional space we have

$$\Delta_L U(x(t)) = \frac{1}{b-a} \int_a^b \frac{\delta^2 U(x)}{\delta x(s)^2} \, ds,$$

where $\delta^2 U(x)/\delta x(s)^2$ is the second-order variational derivative of $U(x(t))$.

But already, in 1914, Volterra [154] had used different second-order differential expressions such as

$$\Delta_0 V(x(t)) = \int_a^b \frac{\delta^2 V(x)}{\delta x(s)\delta x(s)}\, ds$$

(the Volterra Laplacian), where $\delta^2 V(x)/\delta x(s)\delta x(\tau)$ is the second mixed variational derivative of $V(x(t))$. In 1966 Gross [70] and Dalecky [27] independently defined the infinite-dimensional elliptic operator of the second order which includes the Laplace operator

$$\Delta_0 V(x(t)) = \operatorname{Tr} V''(x),$$

where $V''(x)$ is the Hessian of the function $V(x)$ at the point x. For a function V defined on a functional space, $\Delta_0 V(x(t))$ is the Volterra Laplacian, and for functions defined on the space of a countable number of variables, we have

$$\Delta_0 V(x_1, \ldots, x_n, \ldots) = \sum_{k=1}^{\infty} \frac{\partial^2 V}{\partial x_k^2}.$$

There exists a number of other examples of second-order infinite-dimensional differential expressions which considerably differ from the differential expressions of Lévy type. The corresponding references can be found in the bibliography to the monographs of Berezansky and Kondratiev [12] and Dalecky and Fomin [27].

The present book deals with the problems of the theory of equations with the Lévy Laplacians and Lévy–Laplace operators. It is based on the author's papers [36–38, 40, 50–66] and the paper [8].

In Chapter 1 we give the definition of the Lévy Laplacian and describe some of its properties.

In the foreword to his book [95], Lévy wrote: 'In the theories which we mentioned, we essentially face the laws of great numbers similar to the laws of the theory of probabilities . . .'. The probabilistic treatment of the Lévy Laplacian in the second, third, and fourth chapters allows us to enlarge on a number of its interesting properties. Let us mention some of them. The Lévy Laplacian gives rise to operators of arbitrary order depending on the choice of the domain of definition of the operator. There is a huge number of harmonic functions of infinitely many variables connected with the Lévy Laplacian. The natural domain of definition of the Lévy Laplacian and that of the symmetrized Lévy Laplacian do not intersect. Starting from the non-symmetrized Lévy Laplacian, one can construct a symmetric and even a self-adjoint operator.

Problems in the theory of equations with Lévy Laplacians are considered in Chapters 5–7.

First, we concentrate our attention on the main classes of linear elliptic and parabolic equations with Lévy Laplacians.

The equations which describe real physical processes are, as a rule, nonlinear. The theory of linear equations with the Lévy Laplacian is quite developed (see the bibliography). On the other hand, the theory of nonlinear equations with the Lévy Laplacian has only recently begun to be developed. The final two chapters deal with elliptic quasilinear and nonlinear and parabolic nonlinear equations with the Lévy Laplacian.

We will see how striking is the difference (especially in the nonlinear case) between the theories of infinite-dimensional and n-dimensional partial differential equations.

Finally in the Appendix we apply the results of Chapter 3 to the construction of Dirichlet forms associated with the Lévy–Laplace operator, and show the connection between these forms and Markov processes.

There is no doubt that the reader of this book will see that the properties of the Lévy Laplacian, as a rule, have no analogues with the classical finite-dimensional Laplacian. Moreover, the differences are so essential that one can call them pathological if the properties of the Laplace operator for functions of a finite number of variables are considered to be the norm. However, from another point of view the opposite statement is true as well.

It should be emphasized that in this book we consider only the Lévy Laplacian. We do not consider here the problems of the theory of equations and operators of Lévy type (which naturally generalize the equations with Lévy Laplacians and Lévy–Laplace operators) considered in our papers [39, 41–49].

Unfortunately, a lot of the results concerning different trends originated in the book by Lévy are not included in this work although they undoubtedly deserve to be considered. In particular we do not discuss here the well-known approach to the Lévy Laplacian via white noise theory [80, 88]. I hope that this is compensated for to some extent by the large bibliography presented here.

With great warmth I recollect numerous conversations on the topics discussed in this book with those who have departed: Yu. L. Dalecky (1926–1997), O. A. Ladyzhenskaya (1922–2004), E. M. Polishchuk (1914–1987) and G. E. Shilov (1917–1975).

During the preparation of this book for publication I was helped by Ya. I. Belopolskaya and I. I. Kovtun, and I am very grateful to them for their help.

1

The Lévy Laplacian

1.1 Definition of the infinite-dimensional Laplacian

Let H be a countably-dimensional real Hilbert space. Consider a scalar function $F(x)$ on H, where $x \in H$.

Lévy introduced the infinite-dimensional differential Laplacian by

$$\Delta_L F(x_0) = 2 \lim_{\varrho \to 0} \frac{\mathfrak{M}_{(x_0,\varrho)} F(x) - F(x_0)}{\varrho^2}. \tag{1.1}$$

This definition assumes that $F(x)$ has the mean value $\mathfrak{M}_{(x_0,\varrho)} F(x)$, for $\varrho < \varrho_0$, and that the limit at the right-hand side of (1.1) exists.

We define the mean value of the function $F(x)$ over the Hilbert sphere $\|x - x_0\|_H^2 = \varrho^2$ as the limit (if it exists) of the mean value, over the n-dimensional sphere, of the function $F(\sum_{k=1}^n x_k f_k) = f(x_1, \ldots, x_n)$, i.e., of the restriction of the function $F(x)$ on the n-dimensional subspace with the basis $\{f_k\}_1^n$, $x_k = (x, f_k)_H$:

$$\mathfrak{M}_{(x_0,\varrho)} F(x) = \lim_{n \to \infty} M_n F(x),$$

$$M_n F(x) = \frac{1}{s_n} \int_{\sum_{k=1}^n (x_k - x_{0k})^2 = \varrho^2} f(x_1, \ldots, x_n) d\sigma_n,$$

where s_n is the area, and $d\sigma_n$ is the element of the n-dimensional sphere surface. In general, the mean value depends on the choice of the basis.

It follows immediately from its definition that the mean value is additive and homogeneous: if there exists $\mathfrak{M}_{(x_o,\varrho)} F_k$, $k = 1, \ldots, m$, then there exists

$$\mathfrak{M}_{(x_o,\varrho)} \left(\sum_{k=1}^m c_k F_k \right) = \sum_{k=1}^m c_k \mathfrak{M}_{(x_0,\varrho)} F_k.$$

The mean value possesses the multiplicative property: if there exists $\mathfrak{M}_{(x_o,\varrho)} F_k$, and the F_k are uniformly continuous in a bounded domain $\Omega \in H$, which contains the sphere $\|x - x_0\|_H^2 = \varrho^2$, then there exists

$$\mathfrak{M}_{(x_0,\varrho)}\left(\prod_{k=1}^{m} F_k\right) = \prod_{k=1}^{m} \mathfrak{M}_{(x_0,\varrho)} F_k.$$

This property follows from the following statement of Lévy. Let function $F(x)$ be uniformly continuous on the sphere $\|x - x_0\|_H^2 = \varrho^2$, and let the average of the function $F(x)$ exist (i.e., $M_n \to M, M = \mathfrak{M}_{(x_0,\varrho)} F$). Then for each $\delta > 0$ we have

$$\lim_{n\to\infty} \frac{1}{s_n} m_n\{x : |f(x_1, \ldots, x_n) - M| > \delta\} = 0$$

(here m_n denotes the Lebesgue measure).

Note that the definition of the Laplacian via mean values is valid for the finite-dimensional case as well.

The definition (1.1) does not assume differentiability of the function $F(x)$. However, if the function $F(x)$ is twice strongly differentiable, then the following representation of the Lévy Laplacian holds.

Lemma 1.1 *Let the function $F(x)$ be twice strongly differentiable in point x_0, and the Laplacian Δ_L exist. Then*

$$\Delta_L F(x_0) = \lim_{n\to\infty} \frac{1}{n} \sum_{k=1}^{n} (F''(x_0) f_k, f_k)_H, \qquad (1.2)$$

where $F''(x_0)$ is the Hessian of the function $F(x)$ in the point x_0, $F''(x_0) \in \{H \to H\}$, and $\{f_k\}_1^\infty$ is some chosen orthonormal basis in H.

Indeed, it follows from the definition of the mean value that $\mathfrak{M}_{(x_0,\varrho)} F(x) = \mathfrak{M} F(x_0 + \varrho h)$, where $\mathfrak{M}\Phi(h)$ is the mean value of the function Φ over the sphere $\|h\|_H^2 = 1$. Therefore, taking into account that for $a \in H$ $\mathfrak{M}(a, h)_H = 0$, because $\frac{1}{s_n} \int_{\sum_{k=1}^{n} h_k^2 = 1} h_k d\sigma_n = 0$, we have

$$\frac{1}{\varrho^2}\{\mathfrak{M}_{(x_0,\varrho)} F(x) - F(x_0)\} = \frac{1}{\varrho^2}\{\mathfrak{M} F(x_0 + \varrho h) - F(x_0)\}$$

$$= \frac{1}{\varrho^2}\left\{\mathfrak{M}\left[(F'(x_0), \varrho h)_H + \frac{1}{2}(F''(x_0)\varrho h, \varrho h)_H + r(x_0, \varrho h)\right]\right\}$$

$$= \frac{1}{\varrho^2}\left\{\overline{\lim_{n\to\infty}} M_n\left[\frac{\varrho^2}{2}(F''(x_0)h, h)_H + r(x_0, \varrho h)\right]\right\}$$

$$\geq \frac{1}{2}\overline{\lim_{n\to\infty}} M_n(F''(x_0)h, h)_H \underline{\lim}_{n\to\infty} \frac{M_n r(x_0, \varrho h)}{\varrho^2}$$

$$\left(\text{and } \frac{r(x_0, \varrho h)}{\|\varrho h\|_H^2} \to 0 \quad \text{as} \quad \|\varrho h\|_H^2 \to 0\right);$$

similarly we have

$$\frac{1}{\varrho^2}\{\mathfrak{M}_{(x_0,\varrho)}F(x) - F(x_0)\} \leq \frac{1}{2}\underline{\lim}_{n\to\infty}M_n(F''(x_0)h, h)_H + \overline{\lim_{n\to\infty}}\frac{M_n r(x_0, \varrho h)}{\varrho^2}.$$

From this we obtain

$$\frac{1}{\varrho^2}\{\mathfrak{M}_{(x_0,\varrho)}F(x) - F(x_0)\} - \varepsilon(\varrho) \leq \frac{1}{2}\underline{\lim}_{n\to\infty}M_n(F''(x_0)h, h)_H$$

$$\leq \frac{1}{2}\overline{\lim_{n\to\infty}}M_n(F''(x_0)h, h))_H \leq \frac{1}{\varrho^2}\{\mathfrak{M}_{(x_0,\varrho)}F(x) - F(x_0)\} + \varepsilon(\varrho),$$

where $\varepsilon(\varrho) = \frac{1}{\varrho^2}\sup_{\|h\|_H^2=1}|r(x_0, \varrho h)|$, $\varepsilon(\varrho) \to 0$ as $\varrho \to 0$.

Therefore, $\Delta_L F(x_0) = \mathfrak{M}(F''(x_0)h, h)_H$.

Taking into consideration that, according to formula of Ostrogradsky,

$$M_n h_k^2 = \frac{1}{s_n}\int\limits_{\sum_{k=1}^n h_k^2=1} h_k^2\, d\sigma_n = \frac{1}{s_n}\int\limits_{\sum_{k=1}^n h_k^2\leq 1} \frac{\partial h_k}{\partial h_k}dh_1\ldots dh_n$$

$$= \frac{v_n}{s_n} = \frac{\pi^{\frac{n}{2}}/\Gamma(\frac{n}{2}+1)}{2\pi^{\frac{n}{2}}/\Gamma(\frac{n}{2})} = \frac{1}{n},$$

$$M_n h_k h_j = \frac{1}{s_n}\int\limits_{\sum_{k=1}^n h_k^2=1} h_k h_j\, d\sigma_n = 0 \quad \text{for} \quad j \neq k,$$

(here $h_k = (h, f_k)_H$, v_n is volume, s_n is the area of surface of the sphere $\sum_{k=1}^n h_k^2 = 1$, $\Gamma(s)$ the gamma function), we obtain that

$$\Delta_L F(x_0) = \mathfrak{M}(F''(x_0)h, h)_H = \lim_{n\to\infty}\frac{1}{n}\sum_{k=1}^n (F''(x_0)f_k, f_k)_H.$$

$\square$

If at the given point x_0 the function $F(x)$ is twice differentiable only with respect to the subspace Y of the space H (i.e., the second differential of the function $F(x)$ at the point x_0 does not exist for all increments $h \in H$, but $d^2F(x_0, y) = (F_Y''(x_0)y, y)_H$ exists for the increments y that form the subspace Y of the space H, and the second derivative of the function $F(x)$ at the point x_0 with respect to the subspace Y is the operator $F_Y''(x_0) \in \{Y \to Y'\}$, where Y' is the space conjugate to Y), then from (1.1) we deduce that

$$\Delta_L F(x_0) = \lim_{n\to\infty}\frac{1}{n}\sum_{k=1}^n (F_Y''(x_0)f_k, f_k)_H, \tag{1.3}$$

provided that the basis $\{f_k\}_1^\infty$ is orthonormal in H and that $f_k \in Y$.

Now we give the formula for the infinite-dimensional Laplacian obtained by Lévy.

Let there be a function

$$F(x) = f(U_1(x), \ldots, U_m(x)),$$

where $f(u_1, \ldots, u_m)$ is a twice continuously differentiable function of m variables in the domain of values $\{U_1(x), \ldots, U_m(x)\}$ in $\mathbb{R}^m$, $U_j(x)$ are some twice strongly differentiable functions, and the $\Delta_L U_j(x)$ exist ($j = 1, \ldots, m$). Then $\Delta_L F(x)$ exists, and

$$\Delta_L F(x) = \sum_{j=1}^{m} \frac{\partial f}{\partial u_j}\Big|_{u_j = U_j(x)} \Delta_L U_j(x). \tag{1.4}$$

Indeed, the second differential of the function $F(x)$ at the point x for increment $h \in H$ is

$$d^2 F(x; h) = (F''(x)h, h)_H = \sum_{i,j=1}^{m} \frac{\partial^2 f}{\partial u_i \partial u_j}\Big|_{u_l = U_l(x)} (U_i'(x), h)_H (U_j'(x), h)_H$$

$$+ \sum_{j=1}^{m} \frac{\partial f}{\partial u_j}\Big|_{u_j = U_j(x)} (U_j''(x)h, h)_H.$$

According to (1.2),

$$\Delta_L F(x) = \sum_{i,j=1}^{m} \frac{\partial^2 f}{\partial u_i \partial u_j}\Big|_{u_l = U_l(x)} \lim_{n \to \infty} \frac{1}{n} \sum_{k=1}^{n} (U_i'(x), f_k)_H (U_j'(x), f_k)_H$$

$$+ \sum_{j=1}^{m} \frac{\partial f}{\partial u_j}\Big|_{u_j = U_j(x)} \lim_{n \to \infty} \frac{1}{n} \sum_{k=1}^{n} (U_j''(x)f_k, f_k)_H.$$

But

$$\lim_{n \to \infty} \frac{1}{n} \sum_{k=1}^{n} (U_i'(x), f_k)_H (U_j'(x), f_k)_H = 0,$$

(because $(U_l'(x), f_k)_H \to 0$ as $k \to \infty$), and

$$\lim_{n \to \infty} \frac{1}{n} \sum_{k=1}^{n} (U_j''(x)f_k, f_k)_H = \Delta_L U_j(x).$$

Therefore,

$$\Delta_L F(x) = \sum_{j=1}^{m} \frac{\partial f}{\partial u_j}\Big|_{u_j = U_j(x)} \Delta_L U_j(x).$$

A series of consequences follows from formula (1.4).

1. If the functions $U_k(x)$ are harmonic in some domain Ω, $k = 1, \ldots, m$, then the function $F(x)$ also is harmonic in Ω.

2. The Lévy Laplacian is a 'differentiation'. It is enough to set $F(x) = U_1(x)U_2(x)$: then

$$\Delta_L[U_1(x)U_2(x)] = \Delta_L U_1(x) \cdot U_2(x) + U_1(x) \cdot \Delta_L U_2(x).$$

3. The Liouville theorem does not hold for harmonic functions of an infinite number of variables, i.e., there exists a function that is not equal to a constant which is harmonic and bounded in the whole space: it is sufficient to put $F(x) = f(U(x))$, where $f(u)$ is a differentiable function in $\mathbb{R}^1$ bounded together with its derivative, $U(x)$, which is a harmonic function in the whole of H. For example, $F(x) = \cos(\alpha, x)_H$, $\alpha \in H$.

1.2 Examples of Laplacians for functions on infinite-dimensional spaces

For functions on a space of sequences, the Lévy Laplacian is an operator with an infinite number of partial derivatives, and for functions on spaces of functions of finitely many variables, the Lévy Laplacian is an operator in variational derivatives.

Example 1.1 Let $H = l_2$ be the space of sequences $\{x_1, \ldots, x_n, \ldots\}$ such that $\sum_{k=1}^{\infty} x_k^2 < \infty$.

If the function $F(x_1, \ldots, x_n, \ldots)$ is twice strongly differentiable, and the Laplacian exists, then its Hessian is the matrix

$$\left\| \frac{\partial^2 F(x)}{\partial x_i \partial x_k} \right\|_{i,k=1}^{\infty}$$

which induces a bounded operator in l_2 :

$$(F''(x)h, h)_{l_2} = \sum_{i,k=1}^{\infty} \frac{\partial^2 F(x)}{\partial x_i \partial x_k} h_i h_k,$$

and (1.2) yields that

$$\Delta_L F(x_1, \ldots, x_n, \ldots) = \lim_{n \to \infty} \frac{1}{n} \sum_{k=1}^{n} \frac{\partial^2 F(x)}{\partial x_k^2}.$$

Example 1.2 Let $H = L_2(0, 1)$ be the space of functions $x(t)$, square integrable on $[0, 1]$.

If the second differential of the twice differentiable function $F(x(t))$ has the form

$$d^2 F(x; h) = \int_0^1 \frac{\delta^2 F(x)}{\delta x(s)^2} h^2(s)\, ds + \int_0^1 \int_0^1 \frac{\delta^2 F(x)}{\delta x(s)\delta x(\tau)} h(s)h(\tau)\, ds d\tau,$$

where the second variational derivative $\delta^2 F(x)/\delta x(s)^2$ and the second mixed variational derivative $\delta^2 F(x)/\delta x(s)\delta x(\tau)$ are continuous with respect to s and s, τ respectively (here $h(t) \in L_2(0, 1)$), then one says that $d^2 F(x; h)$ has normal form [95], and if

$$d^2 F(x; h) = \int\limits_0^1 \int\limits_0^1 \frac{\delta^2 F(x)}{\delta x(s)\delta x(\tau)} h(s)h(\tau)\, ds d\tau,$$

than one says that it has regular form [154].

We denote by $\mathcal{B}$ the set of all uniformly dense (according to the Lévy terminology) bases in $L_2(0, 1)$, i.e. orthonormal bases $\{f_k\}_1^\infty$ in $L_2(0, 1)$, such that

$$\lim_{n\to\infty} (y, \varphi_n)_{L_2(0,1)} = (y, 1)_{L_2(0,1)} \qquad \text{for all } y \in L_2(0, 1),$$

where $\varphi_n(s) = \frac{1}{n} \sum_{k=1}^n f_k^2(s)$.

As has been shown by Polishchuk (in his comments to the Russian translation of [95]), all orthonormal bases which are the eigenfunctions of some Sturm–Liouville problem are uniformly dense.

Let the function $F(x)$ be twice strongly differentiable, and the second differential have normal form. Then

$$\Delta_L F(x(t)) = \int\limits_0^1 \frac{\delta^2 F(x)}{\delta x(s)^2}\, ds$$

for arbitrary basis from $\mathcal{B}$.

Indeed,

$$(F''(x)h, h)_{L_2(0,1)} = \int\limits_0^1 \int\limits_0^1 \left[\delta(s - \tau)\frac{\delta^2 F(x)}{\delta x(s)^2}\, ds + \frac{\delta^2 F(x)}{\delta x(s)\delta x(\tau)} \right] h(s)h(\tau)\, ds d\tau$$

($\delta(s - \tau)$ is the delta function), and, according to (1.2),

$$\Delta_L F(x(t)) = \lim_{n\to\infty} \left[\int\limits_0^1 \frac{\delta^2 F(x)}{\delta x(s)^2} \varphi_n(s)\, ds + \frac{\delta^2 F(x)}{\delta x(s)\delta x(\tau)} \psi_n(s, \tau)\, ds d\tau \right],$$

where $\psi_n(s, \tau) = \frac{1}{n} \sum_{k=1}^n f_k(s)f_k(\tau)$.

But

$$\int\limits_0^1 \int\limits_0^1 \frac{\delta^2 F(x)}{\delta x(s)\delta x(\tau)} \psi_n(s, \tau)\, ds d\tau \to 0 \quad \text{as} \quad n \to \infty,$$

because

$$\left[\int_0^1\int_0^1 \frac{\delta^2 F(x)}{\delta x(s)\delta x(\tau)}\psi_n(s,\tau)\,ds\,d\tau\right]^2$$

$$\leq \int_0^1\int_0^1 \left[\frac{\delta^2 F(x)}{\delta x(s)\delta x(\tau)}\right]^2 ds\,d\tau \int_0^1\int_0^1 \psi_n^2(s,\tau)\,ds\,d\tau,$$

and

$$\int_0^1\int_0^1 \psi_n^2(s,\tau)\,ds\,d\tau = \frac{1}{n^2}\sum_{k=1}^n \|f_k\|_H^4 = \frac{1}{n}.$$

Taking into account that $\{f_k\}_1^\infty \in \mathcal{B}$, we have

$$\Delta_L F(x) = \int_0^1 \frac{\delta^2 F(x)}{\delta x(s)^2}\,ds.$$

It is clear that if $d^2 F(x;h)$ has regular form, then

$$\Delta_L F(x(t)) = 0.$$

Example 1.3 Let $H = L_{2;m}(0,1)$ be the space of vector functions $x(t) = \{x_1(t),\ldots,x_m(t)\}$, with components square integrable on $[0,1]$.

The second differential of the twice differentiable function $F(x_1(t),\ldots,x_m(t))$ is said to have normal form if

$$d^2 F(x;h) = \sum_{i,j=1}^m \Bigg[\int_0^1 \frac{\delta^2 F(x)}{\delta x_i(s)\delta x_j(s)}h_i(s)h_j(s)\,ds$$

$$+ \int_0^1\int_0^1 \frac{\delta_1^2 F(x)}{\delta x_i(s)\delta x_j(\tau)}h_i(s)h_j(\tau)\,ds\,d\tau\Bigg],$$

where the second partial variational derivatives $\delta^2 F(x)/\delta x_i(s)\delta x_j(s)$ and $\delta_1^2 F(x)/\delta x_i(s)\delta x_j(\tau)$ of the function $F(x)$ are continuous with respect to s and s,τ respectively (here $h(t) \in L_{2;m}(0,1)$); and it is said to have regular form if

$$d^2 F(x;h) = \sum_{i,j=1}^m \int_0^1\int_0^1 \frac{\delta_1^2 F(x)}{\delta x_i(s)\delta x_j(\tau)}h_i(s)h_j(\tau)\,ds\,d\tau.$$

If $F(x)$ is twice strongly differentiable, and the second differential has normal form, then

$$\Delta_L F(x_1(t), \ldots, x_m(t)) = \sum_{j=1}^{m} \int_0^1 \frac{\delta^2 F(x)}{\delta x_j(s)^2}\, ds$$

for an arbitrary uniformly dense basis in $L_{2;m}(0, 1)$ (i.e., for the orthonormal basis $\{f_k\}_1^\infty = \{f_{1k}, \ldots, f_{mk}\}_1^\infty$, such that $\{f_{jk}\}_{k=1}^\infty \in \mathcal{B}$, $j = 1, \ldots, m$).

Indeed,

$$(F''(x)h, h)_{L_{2;m}(0,1)} = \sum_{i,j=1}^{m} \int_0^1 \int_0^1 \left[\delta(s - \tau) \frac{\delta^2 F(x)}{\delta x_i(s) \delta x_j(s)} \right.$$

$$\left. + \frac{\delta_1^2 F(x)}{\delta x_i(s) \delta x_j(\tau)} \right] h_i(s) h_j(\tau)\, ds\, d\tau$$

and, according to (1.2),

$$\Delta_L F(x) = \lim_{n \to \infty} \left[\sum_{j=1}^{m} \int_0^1 \frac{\delta^2 F(x)}{\delta x_j(s)^2} \varphi_{jn}(s)\, ds + \sum_{i \neq j} \int_0^1 \frac{\delta^2 F(x)}{\delta x_i(s) \delta x_j(s)} \psi_{ijn}(s, s)\, ds \right.$$

$$\left. + \sum_{i,j=1}^{m} \int_0^1 \int_0^1 \frac{\delta_1^2 F(x)}{\delta x_i(s) \delta x_j(\tau)} \psi_{ijn}(s, \tau)\, ds\, d\tau \right],$$

where

$$\varphi_{jn}(s) = \frac{1}{n} \sum_{k=1}^{n} f_{jk}^2(s), \quad \psi_{ijn}(s, \tau) = \frac{1}{n} \sum_{k=1}^{n} f_{ik}(s) f_{jk}(\tau).$$

But

$$\int_0^1 \int_0^1 \frac{\delta_1^2 F(x)}{\delta x_i(s) \delta x_j(\tau)} \psi_{ijn}(s, \tau)\, ds\, d\tau \to 0,$$

$$\int_0^1 \frac{\delta^2 F(x)}{\delta x_i(s) \delta x_j(s)} \psi_{ijn}(s, s)\, ds \to 0 \ (i \neq j) \quad \text{as} \quad n \to \infty.$$

Taking into account that $\{f_{jk}\}_1^\infty \in \mathcal{B}$, $j = 1, \ldots, m$, we have

$$\Delta_L F(x) = \sum_{j=1}^{m} \int_0^1 \frac{\delta^2 F(x)}{\delta x_j(s)^2}\, ds.$$

If $d^2 F(x; h)$ has regular form, then

$$\Delta_L F(x_1(t), \ldots, x_m(t)) = 0.$$

1.3 Gaussian measures

The simplest measures in a finite-dimensional space which admit the extension to a Hilbert space are Gaussian measures.

We consider a Gaussian measure in a Hilbert space which we need in the second, third and fourth chapters. We also consider a special Gaussian measure, the Wiener measure, which we need in the fifth chapter.

First we define the Gaussian measure.

The pair $\{H, \mathfrak{A}\}$ where H is a Hilbert space and $\mathfrak{A}$ is a σ-algebra of Borel sets from H is called a measurable Hilbert space.

The measure μ whose characteristic functional has the form

$$\chi(y) = \exp\left\{-\frac{1}{2}(Ky, y)_H + (a, y)_H\right\} \qquad (y \in H),$$

where $a \in H$, K is positive operator of a trace class in H, is called a Gaussian measure in Hilbert space $\{H, \mathfrak{A}\}$.

Here a is the mean value, and K is the correlation operator of the measure μ.

The measure is called centred if $a = 0$. In what follows we consider centred Gaussian measures.

It follows from the above expression for the characteristic functional that all finite-dimensional projections μ_P of the Gaussian measure μ are also Gaussian measures in a finite-dimensional space. In addition, the correlation operator of the measure μ_P has the form $K_P = PKP$ (here P is an ortho-projector onto finite-dimensional (m-dimensional) subspace). The density of the measure μ_P with respect to the Lebesgue measure in $\mathbb{R}^m$ is

$$\varrho(x) = \frac{1}{\sqrt{(2\pi)^m \det K_P}} \exp\left\{-\frac{1}{2}(K_P^{-1}x, x)_{\mathbb{R}^m}\right\}.$$

Finally, one should note the following important statement. The Gaussian measure μ is σ-additive if and only if K is a positive operator of trace class in H.

The triple $\{H, \mathfrak{A}, \mu\}$ is called a measure space.

Now we calculate some integrals with respect to a centred Gaussian measure which we need later.

By direct computation we derive

$$\mu(H) = \int_H \mu(dx) = 1.$$

Hence the triple $\{H, \mathfrak{A}, \mu\}$ is a probability space.

Next we derive the expression for an integral of a function of a finite number of linear functionals with respect to a Gaussian measure. Let

$$F(x) = f((x, \varphi_1)_H, \ldots, (x, \varphi_m)_H),$$

where $f(u_1, \ldots, u_m)$ is a measurable function of m variables, and $\varphi_1, \ldots, \varphi_m$ are orthonormalized functions in H. As long as $F(x)$ is a cylindrical function, we have

$$\int_H F(x)\,\mu(dx) = \frac{1}{\sqrt{(2\pi)^m \det \kappa}} \int_{R^m} f(u_1, \ldots, u_m)e^{-\frac{1}{2}\sum_{j,k=1}^m \tau_{jk}u_j u_k}\,du_1 \cdots du_m,$$

where $\kappa = \|(K\varphi_j, \varphi_k)_H\|_{j,k=1}^m$, and τ_{jk} are the elements of the matrix inverse to κ. The existence of the integral at the left-hand side of this equality yields the existence of the integral on the right-hand side and vice versa.

Now we compute the nth order moments

$$m_{y_1,\ldots,y_n} = \int_H \prod_{k=1}^n (x, y_k)_H\,\mu(dx)$$

of the centred Gaussian measure μ.

If n is odd, $n = 2p - 1$, then

$$m_{y_1,\ldots,y_n} = 0.$$

If n is even, $n = 2p$, then

$$m_{y_1,\ldots,y_n} = (-1)^p \frac{\partial^n}{\partial \varepsilon_1 \cdots \partial \varepsilon_n} \int_H \exp\left\{i(x, \sum_{k=1}^n \varepsilon_k y_k)_H\right\}\mu(dx)\Bigg|_{\varepsilon_1 = \cdots = \varepsilon_n = 0}$$

$$= (-1)^p \frac{\partial^n}{\partial \varepsilon^1 \cdots \partial \varepsilon_n} \exp\left\{-\frac{1}{2}\left(K\left[\sum_{k=1}^n \varepsilon_k y_k\right], \sum_{k=1}^n \varepsilon_k y_k\right)_H\right\}\Bigg|_{\varepsilon_1 = \cdots = \varepsilon_n = 0}$$

$$= \sum_{\{t,\tau\}} \prod_{k=1}^p (Ky_{t_k}, y_{\tau_k})_H \qquad (y_1, \ldots, y_n \in H),$$

where $\sum_{\{t,\tau\}}$ is the sum over all possible partitions of a set of numbers $1, 2, \ldots, n$ into p pairs (t_k, t_{τ_k}), without taking into account the order of the pairs.

In particular,

$$m_{y_1 y_2} = (K\,y_1, y_2)_H,$$

$$m_{y_1 y_2 y_3 y_4} = (K\,y_1, y_2)_H\,(K\,y_3, y_4)_H$$
$$+ (K\,y_1, y_3)_H\,(K\,y_2, y_4)_H + (K\,y_1, y_4)_H\,(K\,y_2, y_3)_H.$$

Note that the computation of the integral $\int_H F(x)\,\mu(dx)$ of the function $F(x)$ admitting an approximation by a sequence of cylindrical functions $F_P(x) = F(Px)$ is reduced to the computation of the integral over the finite-dimensional subspace L_P. If in addition some conditions hold which allow us to pass to the limit under the integral sign, we have

$$\int_H F(x)\,\mu(dx) = \lim_{P\to E}\int_H F_P(x)\mu(dx) = \lim_{P\to E}\int_{L_P} F_P(x)\mu_P(dx).$$

If $P = P_N$, $P_N x = \sum_{k=1}^N x_k e_k$, $x_k = (x, e_k)_H$, where $\{e_k\}_1^\infty$ is a canonical basis in H, then $\mu_P(dx)$ looks particularly simple:

$$\mu_P(dx) = (2\pi)^{-N/2}\prod_{k=1}^N \lambda_k e^{-1/2\sum_{k=1}^N \lambda_k^2 x_k^2}\,dx_1\cdots dx_N.$$

The canonical basis in H is the orthobasis which consists of the eigenvectors of the operator $T = K^{-\frac{1}{2}}$ normalized in H; $T e_k = \lambda_k e_k$, λ_k are eigenvalues of the operator T, $k = 1, 2, \ldots$.

Let $F(x) = \|x\|_H^2$. As long as the sequence of non-negative functions $\|P_N x\|_H^2$ converges to $\|x\|_H^2$ monotonically, then one can go to the limit under the integral sign, and

$$\int_H \|x\|_H^2\,\mu(dx) = \lim_{N\to\infty}\sum_{k=1}^N\int_H (x, e_k)_H^2\,\mu(dx) = \sum_{k=1}^\infty (K e_k, e_k)_H = \operatorname{Tr} K.$$

Let $F(x) = (Ax, x)_H$, where A is a bounded operator in H. Using the expressions for the second order moments, we obtain

$$\int_H (P_N A P_N x, x)_H\,\mu(dx) = \sum_{j,k=1}^N (A e_j, e_k)_H\int_H (x, e_j)_H (x, e_k)_H\,\mu(dx)$$

$$= \sum_{j,k=1}^N (A e_j, e_k)_H (K e_j, e_k)_H.$$

We can go to a limit since $(P_N A P_N x, x)_H \le \|A\|\|x\|_H^2$. Therefore

$$\int_H (Ax, x)_H\,\mu(dx) = \sum_{k=1}^\infty (A K e_k, e_k)_H = \operatorname{Tr} A K.$$

In the same way, using the expression for fourth order moments, we get

$$\int_H (Ax, x)_H^2\,\mu(dx) = \left[\operatorname{Tr} A K\right]^2 + 2\operatorname{Tr}(AK)^2.$$

Now consider the special Gaussian measure in $\hat{L}_2(0, 1)$ with zero mean and correlation operator

$$Kx(s) = \frac{1}{2} \int_0^1 \min(t, s) x(s) \, ds - \frac{1}{2} \int_0^1 \int_0^1 \min(\xi, s) x(s) \, ds \, d\xi$$

and show that this is the Wiener measure. Here $\hat{L}_2(0, 1)$ is the space of real functions $x(t)$, square integrable on $[0, 1]$, satisfying the condition $(x, 1)_{L_2(0,1)} = 0$, where $\hat{L}_2(0, 1)$ is a subspace of $L_2(0, 1)$.

Let $C_0(0, 1)$ be the space of real functions $x(t)$, which are continuous on $[0, 1]$ and satisfy the condition $x(0) = 0$.

Wiener defined a measure in the space $C_0(0, 1)$ as follows. Let $0 = t_0 < t_1 < t_2 < \cdots < t_n = 1$ be a partition of the interval $[0, 1]$. The set of functions $y(t) \in C_0(0, 1)$ satisfying the condition $a_k < y_k < b_k$, where $y_k = y(t_k)$, a_k, b_k are numbers $(k = 1, 2, \ldots, n)$ is called a quasi-interval. Consider quasi-intervals or so-called cylindrical sets in $C_0(0, 1)$.

Define the measure of a quasi-interval Q by the formula

$$\mu_W(Q) =$$

$$\frac{1}{\left[\pi^n \prod_{k=1}^n (t_k - t_{k-1}) \right]^{1/2}} \int_{a_1}^{b_1} \cdots \int_{a_n}^{b_n} \exp\left\{ -\sum_{k=1}^n \frac{(y_k - y_{k-1})^2}{t_k - t_{k-1}} \right\} dy_1 \cdots dy_n.$$

This measure admits a countably-additive continuation to a minimal σ-algebra in $C_0(0, 1)$ which contains all cylindrical sets. The measure μ_W is called the Wiener measure. It is evident that the Wiener measure of the whole space $C_0(0, 1)$ is equal to unity. The support of the Wiener measure is the set of functions which satisfy the Hölder conditions with index $\alpha < 1/2$.

Note that for a wide class of functionals $\Phi(y)$ on the space $C_0(0, 1)$ (in particular, for bounded and continuous functionals), Wiener's integral can be calculated as follows. We replace the function $y(t)$ by a broken line $y_n(t)$ with vertices at points $(t_k, y(t_k))$ and denote by $\Phi_n(y)$ the values of the functional Φ for $y_n(t)$:

$$\Phi_n(y(t)) = \Phi(y_n(t)) = \varphi(y_1, \ldots, y_n),$$

where $\varphi(y_1, \ldots, y_n)$ is a function of n variables, $y_k = y(t_k)$. Then

$$\int_{C_0} \Phi(y) \, \mu_W(dy) = \lim_{n \to \infty} \frac{1}{\left[\pi^n \prod_{k=1}^n (t_k - t_{k-1}) \right]^{1/2}}$$

$$\times \int_{\mathbb{R}^n} \varphi(y_1, \ldots, y_n) \exp\left\{ -\sum_{k=1}^n \frac{(y_k - y_{k-1})^2}{t_k - t_{k-1}} \right\} dy_1 \cdots dy_n.$$

In particular, it allows us to derive the formula the Wiener integral of the functional concentrated in m points. If $\Phi(y) = f(y(t_1), \cdots, y(t_m))$, where $t_1 < t_2 < \cdots < t_m$, and $f(y_1, \ldots, y_m)$ is a function of m variables which is integrable in measure in $\mathbb{R}^m$ having the density

$$\varrho(y_1, \ldots, y_m) = \left[\pi^m \prod_{k=1}^{m}(t_k - t_{k-1})\right]^{-1/2} \exp\left\{-\sum_{k=1}^{n} \frac{(y_k - y_{k-1})^2}{t_k - t_{k-1}}\right\},$$

then

$$\int_{C_0(0,1)} \Phi(y)\mu_W(dx) = \int_{\mathbb{R}^m} f(y_1, \ldots, y_m)\varrho(y_1, \ldots, y_m)\,dy_1 \cdots dy_m.$$

Consider in the space $\hat{L}_2(0, 1)$ the Gaussian measure μ with zero mean and the correlation operator

$$Kx(s) = \frac{1}{2} \int_0^1 \min(t, s)x(s)\,ds - \frac{1}{2} \int_0^1 \int_0^1 \min(\xi, s)x(s)\,ds\,d\xi;$$

$K^{-1} = T^2$ is a closure in $\hat{L}_2(0, 1)$ of the expression $-2(d^2x/dt^2)$ on the set of twice differentiable functions satisfying the condition $x'(0) = x'(1) = 0$.

The operator $K \in \{\hat{L}_2(0, 1) \to \hat{L}_2(0, 1)\}$. This is a trace class positive operator. Indeed, let us find the eigenvalues and eigenvectors of the operator K^{-1}. In other words, we find ρ such that the problem

$$-2\frac{d^2x}{dt^2} = \rho x, \quad x'(0) = x'(1) = 0 \qquad (x \in D_{K^{-1}})$$

has a non-trivial solution.

The general solution of the differential equation

$$2\frac{d^2x}{dt^2} + \rho x = 0$$

has the form

$$x(t) = C_1 \cos\sqrt{\frac{\rho}{2}}\,t + C_2 \sin\sqrt{\frac{\rho}{2}}\,t.$$

It follows from $x'(0) = x'(1) = 0$ that

$$C_2 = 0, \quad C_1 \sin\sqrt{\frac{\rho}{2}}\,t = 0,$$

and for $\sqrt{\rho/2} = \pi k$, $k = 1, 2, \ldots$, there exist non-trivial solutions of the problem

$$x_k(t) = C_1 \cos \pi k t.$$

Therefore, the eigenvalues of the operator K^{-1} are

$$\rho_k = 2\pi^2 k^2.$$

Its eigenvectors, normalized in $\hat{L}_2(0, 1)$ (canonical basis in $\hat{L}_2(0, 1)$), are

$$e_k(t) = \sqrt{2} \cos \pi k t \qquad (k = 1, 2, \ldots).$$

The eigenvalues of K are $\nu_k = 1/(2\pi^2 k^2)$. So $\nu_k > 0$, and

$$\mathrm{Tr}\, K = \frac{1}{2\pi^2} \sum_{k=1}^{\infty} \frac{1}{k^2} = \frac{1}{12}$$

hence K is a positive operator of the trace class in H. Note that eigenvalues of the operator $T = K^{-1/2}$ are

$$\lambda_k = \sqrt{2}\,\pi k.$$

The Gaussian measure with such a correlation operator is a countably additive measure in $\hat{L}_2(0, 1)$. With respect to the measure μ, almost all functions of the space $\hat{L}_2(0, 1)$ are continuous and satisfy the Hölder condition with exponent $\alpha < \frac{1}{2}$.

Let $\hat{C}_0(0, 1)$ be the set of all continuous functions on $[0, 1]$ from $\hat{L}_2(0, 1)$, i.e., the set of continuous functions orthogonal to unity: $\hat{C}_0(0, 1)$ has the full measure in $\hat{L}_2(0, 1)$.

Calculate

$$\mu\left(\left\{x \in \hat{C}_0(0, 1) : \alpha < \int_0^1 f(s)\, dx(s) < \beta\right\}\right),$$

where $f(t)$ is a bounded variation function on $[0, 1]$ such that $\|f\|_{L_2(0,1)} = 1$. Let P be an orthoprojector on $\mathbb{R}^N$, i.e. $Px = \sum_{k=1}^{N} x_k e_k$, $x_k = (x, e_k)_{L_2(0,1)}$. Then

$$\mu\left(\left\{x \in \hat{C}_0(0, 1) : \alpha < \int_0^1 f(s)\, dx(s) < \beta\right\}\right)$$

$$= \lim_{\alpha' \to \alpha, \beta' \to \beta} \lim_{P \to E} \mu\left(\left\{x \in \hat{C}_0(0, 1) : \alpha' < \sum_{k=1}^{N} c_k x_k < \beta'\right\}\right),$$

where $c_k = \int_0^1 f(s)\, de_k(s)$, α^l is a decreasing sequence converging to a number α, β^l is an increasing sequence converging to the number β, and

$$\mu\left(\left\{x \in \hat{C}_0(0, 1) : \alpha^l < \sum_{k=1}^N c_k x_k < \beta^l\right\}\right)$$

$$= \frac{\prod_{k=1}^N \lambda_k}{(2\pi)^{\frac{N}{2}}} \int \cdots \int_{\alpha^l < \sum_{k=1}^N c_k x_k < \beta^l} e^{-\frac{1}{2}\sum_{k=1}^N \lambda_k^2 x_k^2}\, dx_1 \cdots dx_N = \frac{1}{\pi^{\frac{1}{2}}} \int_{\alpha_l\left(2\sum_{k=1}^N \frac{c_k^2}{\lambda_k^2}\right)^{-\frac{1}{2}}}^{\beta_l\left(2\sum_{k=1}^N \frac{c_k^2}{\lambda_k^2}\right)^{-\frac{1}{2}}} e^{-\eta^2}\, d\eta.$$

But

$$2\sum_{k=1}^\infty \frac{c_k^2}{\lambda_k^2} = \sum_{k=1}^\infty \left(\int_0^1 f(s)\sqrt{2}\,\sin\pi k s\, ds\right)^2 = \|f\|_{L_2(0,1)}^2,$$

since $\sqrt{2}\,\sin\pi k t$, $k = 1, 2, \ldots$, is a complete orthonormal system in $L_2(0, 1)$. Therefore

$$\mu\left(\left\{x \in \hat{C}_0(0, 1) : \alpha < \int_0^1 f(s)\, dx(s) < \beta\right\}\right) = \frac{1}{\pi^{\frac{1}{2}}} \int_\alpha^\beta e^{-\eta^2}\, d\eta.$$

If $f_1(t), \ldots, f_n(t)$ are orthonormal functions in $L_2(0, 1)$ having bounded variation on $[0, 1]$, then

$$\mu\left(\left\{x \in \hat{C}_0(0, 1) : \alpha_j < \int_0^1 f_j(s)\, dx(s) < \beta_j \ (j = 1, 2, \ldots, n)\right\}\right)$$

$$= \prod_{j=1}^n \left[\frac{1}{\pi^{\frac{1}{2}}} \int_{\alpha_j}^{\beta_j} e^{-\eta_j^2}\, d\eta_j\right].$$

Choose $f_j(t)$ to be functions of this type.

Let $0 = t_0 \leq t_1 \leq \cdots \leq t_n = 1$ be a partition of $[0, 1]$. Put $f_j(t) = \frac{1}{\sqrt{t_j - t_{j-1}}}$ for $t \in [t_{j-1}, t_j]$, $f_j(t) = 0$ for $t \notin [t_{j-1}, t_j]$; it is clear that $(f_k, f_j)_{L_2(0,1)} = \delta_{kj}$.

Then

$$\mu\left(\left\{x(t) \in \hat{C}_0(0, 1) : a_j < x(t_j) - x(t_{j-1}) < b_j \ (j = 1, 2, \ldots, n)\right\}\right)$$

$$= \prod_{j=1}^{n}\left[\frac{1}{\pi^{\frac{1}{2}}} \int_{\frac{a_j}{\sqrt{t_j - t_{j-1}}}}^{\frac{b_j}{\sqrt{t_j - t_{j-1}}}} e^{-\eta_j^2} d\eta_j\right] = \prod_{j=1}^{n} \frac{1}{[\pi(t_j - t_{j-1})]^{\frac{1}{2}}} \int_{a_j}^{b_j} e^{-\frac{\zeta_j^2}{t_j - t_{j-1}}} d\zeta_j,$$

where $a_j = \alpha_j \sqrt{t_j - t_{j-1}}, \quad b_j = \beta_j \sqrt{t_j - t_{j-1}}$.

There exists the one-to-one correspondence

$$y(t) = x(t) - x(0), \qquad x(t) = y(t) - \int_0^1 y(s) \, ds$$

$$(y(t) \in C_0(0, 1), \quad x(t) \in \hat{C}_0(0, 1))$$

between the space $C_0(0, 1)$ and the space $\hat{C}_0(0, 1)$.

The measure in the space $\hat{C}_0(0, 1)$ is transferred by this correspondence into the space $C_0(0, 1)$. In addition the set

$$\left\{x(t) \in \hat{C}_0(0, 1) : a_j < x(t_j) - x(t_{j-1}) < b_j \quad (j = 1, 2, \ldots, n)\right\}$$

is mapped to the set

$$\left\{y(t) \in C_0(0, 1) : a_j < y(t_j) - y(t_{j-1}) < b_j \quad (j = 1, 2, \ldots, n)\right\},$$

and we have

$$\mu\left(\left\{y(t) \in C_0(0, 1) : a_j < y(t_j) - y(t_{j-1}) < b_j \ (j = 1, 2, \ldots, n)\right\}\right)$$

$$= \prod_{j=1}^{n} \frac{1}{[\pi(t_j - t_{j-1})]^{\frac{1}{2}}} \int_{a_j}^{b_j} e^{-\frac{\zeta_j^2}{t_j - t_{j-1}}} d\zeta_j.$$

By the change of variables $\zeta_j = z_j - z_{j-1}$ $(j = 1, \ldots, n)$, we obtain

$$\mu\left(\left\{y(t) \in C_0(0, 1) : a_j < y(t_j) - y(t_{j-1}) < b_j \ (j = 1, 2, \ldots, n)\right\}\right)$$

$$= \frac{1}{\left[\pi^n \prod_{j=1}^{n}(t_j - t_{j-1})\right]^{\frac{1}{2}}} \int_{a_1}^{b_1} \cdots \int_{a_n}^{b_n} \exp\left\{-\sum_{j=1}^{n} \frac{(z_j - z_{j-1})^2}{t_j - t_{j-1}}\right\} dz_1 \cdots dz_n.$$

But this is the classical Wiener measure of a quasi-interval Q in the space $C_0(0, 1)$. For this reason, one also calls the measure introduced in $\hat{L}_2(0, 1)$ a Wiener measure, preserving the same notation μ_W.

To each functional $F(x(t))$ on $\hat{C}_0(0, 1)$ integrable in the Wiener measure corresponds the functional

$$\Phi(y(t)) = F\left(y(t) - \int_0^1 y(s)\,ds\right)$$

on $C_0(0, 1)$, and $\int_{\hat{C}_0(0,1)} F(x)\mu_W(dx) = \int_{C_0(0,1)} \Phi(y)\,\mu_W(dy)$. But $\hat{C}_0(0, 1)$ has the full measure in $\hat{L}_2(0, 1)$. If the functional $\Phi(y)$ integrable in the Wiener measure on $C_0(0, 1)$ corresponds to the functional $F(x)$ on $\hat{L}_2(0, 1)$ then

$$\int_{\hat{L}_2(0,1)} F(x)\mu_W(dx) = \int_{C_0(0,1)} \Phi(y)\,\mu_W(dy).$$

2

Lévy–Laplace operators

Let $\mathfrak{L}_2(H, \mu)$ be the Hilbert space of functions $F(x)$ on H square integrable in Gaussian measure μ with correlation operator K and zero mean value, and K be a positive operator of trace class such that $||x||_H \leq ||K^{-1/2}x||_H$, $x \in D_{K^{-1/2}}$ (here $D_{K^{-1/2}}$ denotes the domain of definition of the operator $K^{-1/2}$), and $||F||^2_{\mathfrak{L}_2(H,\mu)} = \int_H F^2(x)\,\mu(dx)$.

The Lévy Laplacian is essentially infinite-dimensional: if $F(x)$ is a cylindrical twice differentiable function $F(x) = F(Px)$, P is the projection onto m-dimensional subspace, then its Hessian $F''(x)$ is a finite-dimensional (m-dimensional) operator, and

$$\Delta_L F(x) = \lim_{n \to \infty} \frac{1}{n} \sum_{k=1}^{m} (F''(x)f_k, f_k)_H = 0$$

($\{f_k\}_1^\infty$ is an orthonormal basis in H). At the same time, the set $\mathfrak{C}$ of cylindrical functions is everywhere dense in $\mathfrak{L}_2(H, \mu)$. If we now define the operator in $\mathfrak{L}_2(H, \mu)$ with everywhere dense domain of definition D_L by $LU = \Delta_L U$, $D_L = \mathfrak{C}$, then its closure $\bar{L} = 0$. In particular, we get $\bar{L} = 0$ if we define the operator on the well-known orthonormal system of Fourier–Hermite polynomials [25]

$$\Psi_{m_1...m_N}(x) = \prod_{i=1}^{N} H_{m_i}((K^{-1/2}x, f_i)_H)$$

($N = 1, 2, \ldots$; $m_i = 0, 1, 2, \ldots$), where $\{H_m(\xi)\}_0^\infty$ are partially normalized Hermite polynomials, and $\{f_i\}_1^\infty$ is an orthonormal basis in H, $f_i \in H_+$, because these polynomials are cylindrical functions. Hence it seems that the Lévy–Laplace operator is trivial.

In fact, this is not true. There exist linear sets which are everywhere dense in $\mathfrak{L}_2(H, \mu)$ and on the functions from which the Lévy–Laplace operator is a non-trivial operator.

For example, there exists a wide class of functions for which the Lévy Laplacian exists and is independent of the choice of the basis: namely, the class of twice differentiable functions $F(x)$ such that the mean value of $F(x)$ exists and its Hessian $F''(x)$ is a thin operator: an operator is said to be thin if it has the form $\gamma(x)E + S(x)$, where $\gamma(x)$ is a scalar function, E is the identity operator, and $S(x)$ is a compact operator. If $F''(x)$ is a thin operator then $\Delta_L F(x) = 2\gamma(x)$. The set of such functions includes, among others, the Shilov class [132, 134]. A set of functions of type $F(x) = \phi(x, ||x||_H^2)$, where $\phi(x, ||x||_H^2) = \phi(Px, ||x||_H^2)$, P is the projection on $\mathbb{R}^m$, and $\phi(x, \xi)$ are functions which are defined and twice differentiable on $H \times \mathbb{R}^1$ is called the Shilov class. The Shilov class is dense in $\mathfrak{L}_2(H, \mu)$. If $F(x)$ belongs to the Shilov class then $\Delta_L F(x) = 2(\partial\phi/\partial\xi)|_{\xi=||x||_H^2}$.

Later in this chapter we construct a family of, complete in $\mathfrak{L}_2(H, \mu)$, orthogonal polynomial systems such that the Lévy Laplacian exists and does not depend on the choice of the basis. In addition, the Lévy Laplacian preserves the polynomial within the system.

The Lévy Laplacian has some properties of both first and second order differential expressions, as well as other properties which are not true for either the second or the first order. The domain of definition consisting of orthogonal polynomial systems which do not contain $\mathfrak{C}$ determines 'the order' of the operator, if one takes 'the order' to be the value by which the Lévy Laplacian lowers the degree of the polynomial. In the following we shall see a paradoxical property of the Lévy Laplacian: the same differential expression of Lévy–Laplace generates operators of any order depending on the choice of the domain of definition of the operator.

If the domain of definition coincides with a complete orthonormal system of polynomials it is called natural (see, for example, Emch [34]).

2.1 Infinite-dimensional orthogonal polynomials

Construct a, complete in $\mathfrak{L}_2(H, \mu)$, orthonormal system of polynomials such that the image of the Levy Laplacian belongs to the system.

Let

$$H_\alpha \subseteq H_0 \subseteq H_{-\alpha}, \ H_{+1} \equiv H_+, \ H_0 \equiv H, \ H_{-1} \equiv H_-, \ \alpha > 0$$

be a chain of spaces from a Hilbert scale $\{H_\beta\}$, $-\infty < \beta < \infty$, with the generating operator $T = K^{-1/2}$ (its inverse, T^{-1}, is the Hilbert–Schmidt operator),

$$(x, y)_{H_\beta} = (T^\beta x, T^\beta y)_H \quad (x, y \in H_\beta).$$

We denote by $H' \otimes H''$ the tensor product of spaces H' and H''.

A linear combination of homogeneous forms of degrees not higher than m is called a polynomial of degree m on H. A real function $\Phi_m(x) = \underbrace{\varphi(x, \ldots, x)}_{m}$, where $\underbrace{\varphi(x, \ldots, w)}_{m}$ is a symmetric m-linear form $(x, \ldots, w \in H)$, $m = 1, 2, \ldots$, is called a homogeneous form of degree m; the Φ_0 are just numbers.

Let $\hat{\Lambda}_m$ be the set of all measurable polynomials in $\mathfrak{L}_2(H, \mu)$ of degree less than or equal to m. Then $\hat{\Lambda}_m$ is a subspace of $\mathfrak{L}_2(H, \mu)$, and $\hat{\Lambda}_0 \subset \hat{\Lambda}_1 \subset \cdots \subset \hat{\Lambda}_m$. We denote by Λ_m the orthogonal complement in $\hat{\Lambda}_m$ to Λ_{m-1}, i.e. $\hat{\Lambda}_m = \Lambda_m \oplus \hat{\Lambda}_{m-1}$, $m = 1, 2, \ldots$; $\Lambda_0 = \hat{\Lambda}_0$.

The subspaces $\Lambda_0, \Lambda_1, \ldots, \Lambda_m$ are mutually orthogonal, $\hat{\Lambda}_m = \oplus \sum_{k=0}^{m} \Lambda_k$, and, as long as the set of measurable polynomials is dense in $\mathfrak{L}_2(H, \mu)$, we have

$$\mathfrak{L}_2(H, \mu) = \oplus \sum_{k=0}^{\infty} \Lambda_k.$$

To each form $\Phi_m \in \hat{\Lambda}_m$ corresponds its projection onto subspace Λ_m, namely, the polynomial $P_m \in \Lambda_m$.

Now we take, for $\Phi_m(x)$, the form $\Phi_{\gamma_m}(x)$ which represents the measurable extension to H of the continuous form $(m!)^{-1/2}(\gamma_m, \otimes y^m)_{\otimes H^m}$, where the kernel $\gamma_m \in \otimes H_-^m$ is such that $\Phi_{\gamma_m}(x) \in \mathfrak{L}_2(H, \mu)$: $(\gamma_m, y \otimes \ldots \otimes w)_{\otimes H^m}$ is a symmetric m-linear continuous form $(\gamma_m \in \otimes H_-^m, y, \ldots, w \in H_+)$. In the what follows we use the notation

$$\Phi_{\gamma_m}(x) = (m!)^{-1/2}(\gamma_m, \otimes x^m)_{\otimes H^m}$$

for its measurable extension to H (although here $x \in H$, and not H_+).

The projection of such a form $\Phi_{\gamma_m}(x)$ onto Λ_m is a polynomial $P_{\gamma_m}(x)$, which consists of $\Phi_{\gamma_m}(x)$ and the homogeneous forms of smaller degree $\Psi_\nu(x, \gamma_m)$, $\nu < m$. In addition,

$$(P_{\gamma_m}, P_{\gamma_n})_{\mathfrak{L}_2(H,\mu)} = 0, \quad \text{for all} \quad P_{\gamma_m} \in \Lambda_m, \quad \text{for all} \quad P_{\gamma_n} \in \Lambda_n, \quad n < m.$$

To prove the existence of a measurable extension, we show that

$$(P_{\gamma_m}, P_{\sigma_m})_{\mathfrak{L}_2(H,\mu)} = (\gamma_m, \sigma_m)_{\otimes H^m} \quad \text{for all} \quad \gamma_m, \quad \sigma_m \in \otimes H_-^m. \tag{2.1}$$

To this end we calculate the integral of the nth differential in Gaussian measure.

Theorem 2.1 *Let $F(x) \in \mathfrak{L}_2(H, \mu)$, $d^i F(x; h_1, \ldots, h_i)$, $i = 1, \ldots, n$, exist for arbitrary $h_1, \ldots, h_n \in H_{+2}$, (i.e., $F(x)$ is differentiable with respect to the subspace H_{+2}) and there exists $\varepsilon_i > 0$ such that*

$$\sup_{|\varepsilon| \leq \varepsilon_i} |d^i F(x + \varepsilon h_i; h_1, \ldots, h_i)| \, \forall \prod_{j=i+1}^{n} [1 + (h_j, x)_{H_+}] \in \mathfrak{L}(H, \mu), i = 1, \ldots, n.$$

Then

$$\int_H d^n F(x; h_1, \ldots, h_n) \mu(dx)$$

$$= \sum_{j=0}^{[n/2]} (-1)^j \Big[\sum_{\{n,j;t,\tau,s\}} \prod_{k=1}^{j} (h_{i_k}, h_{\tau_k})_{H_+} \int_H F(x) \prod_{i=1}^{n-2j} (h_{s_i}, x)_{H_+} \mu(dx) \Big], \quad (2.2)$$

where $\sum_{\{n,j;t,\tau,s\}}$ is the sum over all possible combinations of j pairs of numbers $\{t_1, \tau_1\}, \ldots, \{t_j, \tau_j\}$ chosen from numbers $1, 2, \ldots, n$ ($s_1, \ldots, s_{n-2j}$ are the remaining numbers), without taking into account the order of the j pairs and of the remaining $n - 2j$ numbers.

Proof. By the shift transformation, since we can pass to the limit, and differentiation under the integral sign for $n = 1$ we have

$$\int_H d F(x; h_1) \mu(dx) = \int_H \frac{d}{d\varepsilon_1} F(x + \varepsilon h_1) \Big|_{\varepsilon_1 = 0} \mu(dx)$$

$$= \frac{d}{d\varepsilon_1} e^{-\frac{\varepsilon_1^2}{2} \|h_1\|_{H_+}^2} \int_H F(x) e^{\varepsilon_1 (h_1, x)_{H_+}} \mu(dx) \Big|_{\varepsilon_1 = 0}$$

$$= \int_H F(x)(h_1, x)_{H_+} \mu(dx). \qquad (2.3)$$

The rest we prove by induction. Assume that formula (2.2) is valid for n; we show that it is valid for $n + 1$ as well:

$$\int_H d^{n+1} F(x; h_1, \ldots, h_{n+1}) \mu(dx) = \sum_{j=0}^{[n/2]} (-1)^j \Big[\sum_{\{n,j;t,\tau,s\}} \prod_{k=1}^{j} (h_{t_k}, h_{\tau_k})_{H_+}$$

$$\times \Big(\int_H d \Big\{ F(x) \prod_{i=1}^{n-2j} (h_{s_i}, x)_{H_+}; h_1 \Big\} \mu(dx)$$

$$- \int_H F(x) \sum_{l=1}^{n-2j} (h_{s_l}, h_1)_{H_+} \prod_{i=1, i \neq l}^{n-2j} (h_{s_i}, x)_{H_+} \mu(dx) \Big) \Big],$$

where $t_k, \tau_k, s_i = 2, 3, \ldots, n + 1$.

It follows from (2.3) and the theorem's conditions that

$$\int_H d^{n+1} F(x; h_1, \ldots, h_{n+1}) \mu(dx) = \sum_{j=0}^{[\frac{n+1}{2}]} (-1)^j \Big[\sum_{\{n,j;t,\tau,s\}} \prod_{k=1}^{j} (h_{t_k}, h_{\tau_k})_{H_+}$$

$$\times \int_H F(x) \prod_{i=1}^{n-2j} (h_{s_i}, x)_{H_+} \mu(dx) \Big]. \qquad \square$$

Let us show that (2.1) holds. Since Λ_m and $\hat{\Lambda}_\nu$ are orthogonal if $\nu < m$, we have

$$(P_{\gamma_m}, P_{\sigma_m})_{\mathcal{L}_2(H,\mu)} = \int_H P_{\gamma_m}(x)\left[\Phi_{\sigma_m}(x) + \sum_{\nu=1}^{m-1}\Psi_\nu(x)\right]\mu(dx)$$

$$= \int_H P_{\gamma_m}(x)\Phi_{\sigma_m}(x)\mu(dx).$$

It can be shown from Theorem 2.1 that

$$(P_{\gamma_m}, P_{\sigma_m})_{\mathcal{L}_2(H,\mu)} = \int_H \Phi_{\sigma_m}(x)P_{\gamma_m}(x)\mu(dx)$$

$$= \int_H \sum_{\nu_1,\ldots,\nu_m=1}^{\infty} (m!)^{-\frac{1}{2}}(\sigma_m, f_{\nu_1}\otimes\ldots\otimes f_{\nu_m})_{\otimes H^m}$$

$$\times (x, f_{\nu_1})_H \ldots (x, f_{\nu_m})_H \, P_{\gamma_m}(x)\mu(dx)$$

$$= \int_H \sum_{\nu_1,\ldots,\nu_m=1}^{\infty} (m!)^{-\frac{1}{2}}(\sigma_m, f_{\nu_1}\otimes\ldots\otimes f_{\nu_m})_{\otimes H^m}$$

$$\times\left\{d^m P_{\gamma_m}(x; T^{-2}f_{\nu_1},\ldots,T^{-2}f_{\nu_m})\right.$$

$$-\sum_{j=1}^{[\frac{m}{2}]}(-1)^j\left[\sum_{\{n,j;t,\tau,s\}}\prod_{k=1}^{j}(f_{t_k}, f_{\tau_k})_{H_-}\right.$$

$$\left.\left.\times\prod_{i=1}^{m-2j}(f_{s_i}, x)_H P_{\gamma_m}(x)\right]\right\}\mu(dx).$$

Since

$$\prod_{i=1}^{m-2j}(f_{s_i}, x)_H \in \hat{\Lambda}_{m-2j}, \qquad P_{\gamma_m}\in\Lambda$$

we get

$$\int_H \prod_{i=1}^{m-2j}(f_{s_i}, x)_H P_{\gamma_m}(x)\mu(dx) = 0, \qquad j = 1,\ldots,\left[\frac{m}{2}\right].$$

Therefore

$$(P_{\gamma_m}, P_{\sigma_m})_{\mathcal{L}_2(H,\mu)} = \sum_{\nu_1,\ldots,\nu_m=1}^{\infty}(\sigma_m, f_{\nu_1}\otimes\ldots\otimes f_{\nu_m})_{\otimes H^m}$$

$$\times (\gamma_m, T^{-2}f_{\nu_1},\ldots,T^{-2}f_{\nu_m})_{\otimes H^m} = (\gamma_m, \sigma_m)_{\otimes H_-^m};$$

here $\{f_k\}_1^{\infty}$ is an orthonormal basis in H.

Let us show the existence of the measurable extension. Let $\gamma_{m,k}$ converge to γ_m in $\otimes H_-^m$, $\gamma_{m,k} \in \otimes H^m$, $\gamma_m \in \otimes H_-^m$. Then (2.1) shows that the sequence $P_{\gamma_{m,k}}(x)$ converges in $\mathfrak{L}_2(H, \mu)$. Choose a subsequence $P_{\gamma_{m,k_i}}(x)$ (the $P_{\gamma_{m,k_i}}(x)$ are continuous) which converges almost everywhere on H, and put $P_{\gamma_m}(x) = \lim_{i \to \infty} P_{\gamma_{m,k_i}}(x)$. Using (2.1) it can be shown that $P_{\gamma_m}(x)$ is unique.

Finally, as long as for all $\gamma_m \in \otimes H_-^m$ there exists a sequence $\gamma_{m,k} \in \otimes H^m$, which converges to γ_m in $\otimes H_-^m$, we have $(P_{\gamma_m}, P_{\sigma_m})_{\mathfrak{L}_2(H,\mu)} = (\gamma_m, \sigma_m)_{\otimes H_-^m}$, for all $\gamma_m, \sigma_m \in \otimes H_-^m$.

Wiener [157] constructed the system of orthogonal polynomials for the case of Wiener measure for the functionals defined on the space of continuous functions. We use this method to give the explicit form of polynomials by orthogonalizing the sequence of forms (more exactly, of the classes of forms)

$$\Phi_{\gamma_m}(x) = (m!)^{-1/2}(\gamma_m, \otimes x^m)_{\otimes H^m}, \quad \gamma_m \in \otimes H_-^m.$$

Each polynomial $P_{\gamma_m}(x)$ consists of $\Phi_{\gamma_m}(x)$ and of the forms of smaller degree $\Psi_\nu(x, \gamma_m)$, $\nu < m$:

$$P_0 = 1, \quad P_{\gamma_1}(x) = \Phi_{\gamma_1}(x),$$

$$P_{\gamma_m}(x) = \Phi_{\gamma_m}(x) + \sum_{k=1}^{[\frac{m}{2}]} \Psi_{m-2k}(x, \gamma_m) \quad (m = 2, 3, \ldots).$$

We construct $\Psi_{m-2k}(x, \gamma_m)$ to obtain $(P_{\gamma_m}, P_{\gamma_{m-2k}})_{\mathfrak{L}_2(H,\mu)} = 0$ given $\Phi_{\gamma_m}(x)$ $(k = 1, \ldots, [m/2])$.

Since the forms of even and odd degrees are orthogonal, we have

$$(P_{\gamma_{2p}}, P_{\gamma_{2q+1}})_{\mathfrak{L}_2(H,\mu)} = 0 \quad (p, q = 0, 1, \ldots).$$

For $P_{\gamma_2}(x) = \Phi_{\gamma_2}(x) + \Psi_0(\gamma_2)$, it follows from

$$(P_{\gamma_2}, P_0)_{\mathfrak{L}_2(H,\mu)} = \int_H [\Phi_{\gamma_2}(x) + \Psi_0(\gamma_2)]\mu(dx) = 0$$

that

$$\Psi_0(\gamma_2) = \int_H \Phi_{\gamma_2}(x)\mu(dx).$$

For $P_{\gamma_3}(x) = \Phi_{\gamma_3}(x) + \Psi_1(x, \gamma_3)$, and using the results from $(P_{\gamma_3}, P_{\gamma_1})_{\mathfrak{L}_2(H,\mu)} = 0$ we have $\Psi_1(x, \gamma_1) = (\sigma_1, x)_H$. According to Theorem 2.1, for $n = 1$ we have

$$\int_H dF(x; h)\mu(dx) = \int_H F(x)(h, x)_{H_+}\mu(dx).$$

Therefore, taking into account that $P_{\gamma_1}(x) = \sum_{i=1}^{\infty} (\gamma, f_i)_H (x, f_i)_H$, we have

$$\int_H (x, f_i)_H P_{\gamma_3}(x)\mu(dx) = \int_H d P_{\gamma_3}(x; T^{-2} f_i)\mu(dx)$$

$$= \int_H d\Phi_{\gamma_3}(x; T^{-2} f_i)\mu(dx) + (\sigma_1, T^{-2} f_i)_H = 0.$$

Hence

$$(\sigma_1, T^{-2} f_i)_H = -\int_H d\Phi_{\gamma_3}(y; T^{-2} f_i)\mu(dy)$$

$$= -\frac{3}{\sqrt{3!}} \int_H (\gamma_3, T^{-2} f_i \otimes y \otimes y)_{\otimes H^3}\mu(dy),$$

and

$$\Psi_1(x, \gamma_3) = -\frac{3}{\sqrt{3!}} \int_H (\gamma_3, x \otimes y \otimes y)_{\otimes H^3}\mu(dy).$$

For $P_{\gamma_4}(x) = \Phi_{\gamma_4}(x) + \Psi_2(x, \gamma_4) + \Psi_0(\gamma_4)$, from the condition $(P_{\gamma_4}, P_0)_{\mathfrak{L}_2(H,\mu)} = 0$, we find that

$$\Psi_0(\gamma_4) = -\int_H \Phi_{\gamma_4}(x) + \Psi_2(x, \gamma_4)]\mu(dx).$$

From the condition $(P_{\gamma_4}, P_{\gamma_2})_{\mathfrak{L}_2(H,\mu)} = 0$ we find

$$\Psi_2(x, \gamma_4) = (\sigma_2, x \otimes x)_{H \otimes H}.$$

According to Theorem 2.1, for $n = 2$ we have

$$\int_H d^2 F(x; h_1, h_2)\mu(dx) = \int_H F(x)(x, h_1)_{H_+}(x, h_2)_{H_+}\mu(dx)$$

$$-(h_1, h_2)_{H_+}\int_H F(x)\mu(dx).$$

Therefore, taking into account that $(P_{\gamma_4}, P_0)_{\mathfrak{L}_2(H,\mu)} = 0$ and that

$$P_{\gamma_2}(x) = \sum_{j,k=1}^{\infty} (\gamma_2, f_j \otimes f_k)_{H \otimes H}(x, f_j)_H (x, f_k)_H,$$

we have

$$\int_H (x, f_i)_H (x, f_k)_H P_{\gamma_4}(x)\mu(dx) = \int_H d^2 P_{\gamma_4}(x; T^{-2} f_j, T^{-2} f_k)\mu(dx)$$

$$= \int_H d^2 \Phi_{\gamma_4}(x; T^{-2} f_j, T^{-2} f_k)\mu(dx) + 2(\sigma_2, T^{-2} f_j \otimes T^{-2} f_k)_{H \otimes H} = 0.$$

Hence

$$(\sigma_2, T^{-2}f_j \otimes T^{-2}f_k)_{H \otimes H} = -\frac{1}{2}\int_H d^2\Phi_{\gamma_4}(y; T^{-2}f_j, T^{-2}f_k)\mu(dy)$$

$$= -\sqrt{\frac{3}{2}}\int_H (\gamma_4, T^{-2}f_j \otimes T^{-2}f_k \otimes y \otimes y)_{\otimes H^4}\mu(dy)$$

and

$$\Psi_2(x, \gamma_4) = -\sqrt{\frac{3}{2}}\int_H (\gamma_4, x \otimes x \otimes y \otimes y)_{\otimes H^4}\mu(dy).$$

Continuing in a similar way, we construct the following system of orthogonal polynomials:

$$P_0(x) = 1, \quad P_{\gamma_1}(x) = (\gamma_1, x)_H,$$

$$P_{\gamma_2}(x) = \frac{1}{\sqrt{2!}}(\gamma_2, x \otimes x)_{H \otimes H} - \frac{1}{\sqrt{2!}}\int_H (\gamma_2, x \otimes x)_{H \otimes H}\mu(dx),$$

$$P_{\gamma_3}(x) = \frac{1}{\sqrt{3!}}(\gamma_3, \otimes x^3)_{\otimes H^3} - \frac{3}{\sqrt{3!}}\int_H (\gamma_2, x \otimes y^2)_{\otimes H^3}\mu(dy),$$

$$P_{\gamma_4}(x) = \frac{1}{\sqrt{4!}}(\gamma_4, \otimes x^4)_{\otimes H^4} - \frac{6}{\sqrt{4!}}\int_H (\gamma_4, \otimes x^2 \otimes y^2)_{\otimes H^4}\mu(dy)$$

$$+ \frac{6}{\sqrt{4!}}\int_H \int_H (\gamma_4, \otimes x^2 \otimes y^2)_{\otimes H^4}\mu(dy)\mu(dx)$$

$$- \frac{1}{\sqrt{4!}}\int_H (\gamma_4, \otimes x^4)_{\otimes H^4}\mu(dx),$$

etc.

Let $\{\gamma_{mq}\}_{q=1}^{\infty}$ be a complete orthonormal system in $\otimes H_-^m$.

Lemma 2.1 *The system of polynomials $P_0 = 1$, $P_{\gamma_{mq}}$, with $m, q = 1, 2, \ldots$, makes an orthonormal basis in $\mathfrak{L}_2(H, \mu)$.*

Indeed, $(P_{\gamma_{mq}}, P_{\gamma_{np}})_{\mathfrak{L}_2(H,\mu)} = 0$ for $m \neq n$, and, according to (2.1), $(P_{\gamma_{mq}}, P_{\gamma_{mp}})_{\mathfrak{L}_2(H,\mu)} = (\gamma_{mq}, \gamma_{mp})_{\otimes H_-^m} = \delta_{pq}$ (here δ_{pq} is the Kronecker delta); therefore

$$(P_{\gamma_{mq}}, P_{\gamma_{np}})_{\mathfrak{L}_2(H,\mu)} = \delta_{mn}\delta_{pq}.$$

The completeness of the system P_0, $P_{\gamma_{mq}}$ in $\mathfrak{L}_2(H, \mu)$ follows from the fact that $P_{\gamma_{mq}} \in \Lambda_m$, $\mathfrak{L}_2(H, \mu) = \oplus \sum_{m=0}^{\infty} \Lambda_m$, while the systems $\{\gamma_{mq}\}_{q=1}^{\infty}$ are complete in $\otimes H_-^m$, $m = 1, 2, \ldots$. $\square$

As usual it follows from Lemma 2.1 that given $U \in \mathfrak{L}_2(H, \mu)$ we have its expansion

$$U(x) = U_0 P_0 + \sum_{m,q=1}^{\infty} U_{mq} P_{\gamma_{mq}}(x),$$

where

$$U_0 = \int_H U(x)\mu(dx), \quad U_{mq} = \int_H U(x)P_{\gamma_{mq}}\mu(dx).$$

This series converges to $U(x)$ in $\mathfrak{L}_2(H, \mu)$, and

$$||U||_{\mathfrak{L}_2(H,\mu)}^2 = U_0^2 + \sum_{m,q}^{\infty} U_{mq}^2.$$

2.2 The second-order differential operators generated by the Lévy Laplacian

Consider operators of the second order generated by the differential expression of Lévy–Laplace.

Choose kernels such that forms corresponding to them contain $||x||_H^{2l}$. To this end we use the kernel corresponding to the identity operator E in H. Due to the theorem of Berezansky about a kernel [11]

$$(Ey, z)_H = (\delta, y \otimes z)_{H_- \otimes H_-}, \quad \delta \in H_- \otimes H_-, y, z \in H_+$$

and

$$||x||_H^{2l} = (\otimes \delta^l, \otimes x^{2l})_{\otimes H^{2l}}.$$

By (1.4), for $m = 1$ we have

$$\Delta_L[||x||_H^2]^l = l[||x||_H^2]^{l-1} \Delta_L ||x||_H^2.$$

Since $([||x||_H^2]''h, h)_H = 2(Eh, h)_H$, by (1.2), we have

$$\Delta_L ||x||_H^2 = \lim_{n \to \infty} \frac{1}{n} \underbrace{(2 + \cdots + 2)}_{n} = 2.$$

Therefore

$$\Delta_L[||x||_H]^{2l} = 2l ||x||_H^{2(l-1)},$$

i.e., the Lévy Laplacian decreases the degree of the form $(\otimes \delta^l, \otimes x^{2l})_{\otimes H^{2l}}$ by 2.

Put $\gamma_{mq} = a_{mq}$, where $\{a_{mq}\}_{q=1}^{\infty}$ is a complete orthonormal system of elements in $\otimes H_{-}^{m}$ such that

$$a_{2n,q} = \sum_{p=1}^{n} \frac{\mu_{2n,q}\mu_{2n-2,q}\cdots\mu_{2p,q}[\otimes\delta^{n-p+1}\tilde{\otimes}s_{2p-2,q}]}{(n-p+1)!} + \tilde{s}_{2n,q},$$

$$a_{2n-1,q} = \sum_{p=1}^{n} \frac{\mu_{2n-1,q}\mu_{2n-3,q}\cdots\mu_{2p-1,q}[\otimes\delta^{n-p}\tilde{\otimes}s_{2p-1,q}]}{(n-p)!}, \quad (n = 1, 2, \ldots).$$

$$(2.4)$$

Here $\{s_{mq}\}_{q=1}^{\infty}$ is a complete sequence of elements in $\otimes H_{-}^{m}$ such that $s_{mq} \in \otimes H_{+}^{m}$, and $\tilde{s}$ denotes that the symmetrization is applied to s. The numbers μ_{mq} are obtained in when b_{mq} is orthogonalized ($q = 1, 2, \ldots$). The system $a_{2,q} = \mu_{2,q}\delta + \tilde{s}_{2,q}$ is obtained when the sequence $b_{2,q} = \delta + t_{2,q}$ is orthogonalized, the system $a_{4,q} = \frac{1}{2}\mu_{4,q}\mu_{2,q}\delta\tilde{\otimes}\delta + \mu_{4,q}\delta\tilde{\otimes}s_{2,q} + \tilde{s}_{4,q}$ is obtained when the sequence $b_{4,q} = \mu_{2,q}\delta\tilde{\otimes}\delta + s_{2,q}\tilde{\otimes}\delta + \tilde{t}_{4,q}$ is orthogonalized and so on, where $\{t_{mq}\}_{q=1}^{\infty}$ is a complete sequence in $\otimes H_{-}^{m}$, $t_{m,q} \in \otimes H_{+}^{m}$. The completeness of the system $\{a_{mq}\}_{q=1}^{\infty}$ in $\otimes H_{-}^{m}$ follows from the completeness of the sequence $\{s_{mq}\}_{q=1}^{\infty}$.

Denote $\mathcal{P}_{a_{mq}} \equiv \mathcal{P}_{mq}$.

According to Lemma 2.1, the system of polynomials $\mathcal{P}_0$, $\mathcal{P}_{mq}(x)$, for $m, q = 1, 2, \ldots$, makes an orthonormal basis in $\mathfrak{L}_2(H, \mu)$.

We denote by $\mathfrak{P}$ the set of all possible linear combinations $A_0\mathcal{P}_0 + \sum_{m,q=1}^{N} A_{mq}\mathcal{P}_{mq}$, where A_{mq} are arbitrary numbers, and N is a natural number.

Theorem 2.2 *The Lévy Laplacian on* $\mathfrak{P}$ *exists, does not depend on the choice of the basis, and is a second order operator. It decreases the degree of the polynomial* $\mathcal{P}_{mq}(m \geq 2)$ *by 2:*

$$\Delta_L \mathcal{P}_0 = 0, \quad \Delta_L \mathcal{P}_{1q}(x) = 0,$$

$$\Delta_L \mathcal{P}_{mq}(x) = 2[m(m-1)]^{-1/2}\mu_{mq}\mathcal{P}_{m-2,q}(x) \quad (m = 2, 3, \ldots).$$

Proof. According to (2.4), we have

$$\Phi_{a_{2n,q}}(x) = (2n!)^{-\frac{1}{2}}(a_{2n,q}, \otimes x^{2n})_{\otimes H^{2n}}$$

$$= (2n!)^{-\frac{1}{2}}\Big[\sum_{p=1}^{n} \frac{\mu_{2n,q}\cdots\mu_{2p,q}\|x\|_{H}^{2n-2p+2}(\tilde{s}_{2p-2,q}, \otimes x^{2p-2})_{\otimes H^{2p-2}}}{(n-p+1)!}$$

$$+ (\tilde{s}_{2n,q}, \otimes x^{2n})_{\otimes H^{2n}}\Big].$$

The forms $\Phi_{a_{2n,q}}(x) \in \mathfrak{L}_2(H, \mu)$, since $\int_H ||x||_H^{4l} \mu(dx) < \infty$. The integral $\int_H ||x||_H^{4l} \mu(dx)$ can be easily computed using the equality

$$\int_H (x, e_k)_H^{2p} \mu(dx) = \frac{\lambda_k}{\sqrt{2\pi}} \int_{-\infty}^{\infty} x_k^{2p} \exp\{-\lambda_k^2 x_k^2\} dx_k = \frac{(2p-1)!!}{\lambda_k^{2p}},$$

where $\{e_k\}_1^{\infty}$ is a canonical basis in H.

An orthogonal basis $\{e_k\}_1^{\infty}$ in H is called canonical if e_k are eigenvectors of the operator T, normalized in H, i.e. $T e_k = \lambda_k e_k$ and λ_k are eigenvalues of the operator T, $k = 1, 2, \ldots$.

For example,

$$\int_H ||x||_H^4 \mu(dx) = 2 \sum_{k=1}^{\infty} \frac{1}{\lambda_k^4} + \left(\sum_{k=1}^{\infty} \frac{1}{\lambda_k^2} \right)^2 = 2\mathrm{Tr}K^2 + (\mathrm{Tr}K)^2.$$

For functions

$$H_{lv}(x) = \frac{1}{l!} ||x||_H^{2l} (\tilde{s}_{vq}, \otimes x^v)_{\otimes H^v},$$

by Corollary 2 to (1.4), we obtain

$$\Delta_L H_{lv}(x) = \Delta_L \left[\frac{1}{l!} ||x||_H^{2l} \right] \cdot (\tilde{s}_{vq}, \otimes x^v)_{\otimes H^v} + \frac{1}{l!} ||x||_H^{2l} \cdot \Delta_L (\tilde{s}_{vq}, \otimes x^v)_{\otimes H^v}.$$

Here

$$\Delta_L \left[\frac{1}{l!} ||x||_H^{2l} \right] = \frac{2}{(l-1)!} ||x||_H^{2(l-1)},$$

and by (1.2),

$$\Delta_L (\tilde{s}_{vq}, \otimes x^v)_{\otimes H^v} = v(v-1) \lim_{n \to \infty} \frac{1}{n} \sum_{k=1}^{n} (\tilde{s}_{vq}, \otimes x^{v-2} \otimes f_k \otimes f_k)_{\otimes H^v} = 0;$$

note $(\tilde{s}_{vq}, \otimes x^{v-2} \otimes f_k \otimes f_k)_{\otimes H^v} \to 0$ as $k \to \infty$, since $\tilde{s}_{vq} \in \otimes H_+^v$. Thus we have

$$\Delta_L H_{lv}(x) = \frac{2}{(l-1)!} ||x||_H^{2(l-1)} (\tilde{s}_{vq}, \otimes x^v)_{\otimes H^v}$$

for an arbitrary basis $\{f_k\}_1^{\infty}$ in H.

Hence

$$\Delta_L \Phi_{a_{2n,q}}(x) = 2\mu_{2n,q} [2n(2n-1)]^{-1/2} \Phi_{a_{2n-2,q}}(x).$$

If $m = 2n + 1$, then similarly we have

$$\Delta_L \Phi_{a_{2n+1,q}}(x) = 2\mu_{2n+1,q} [(2n+1)2n]^{-1/2} \Phi_{a_{2n-1,q}}(x).$$

As the result we obtain

$$\Delta_L \mathcal{P}_{mq}(x) = 2\mu_{mq} [m(m-1)]^{-1/2} \mathcal{P}_{m-2,q}(x).$$

$\square$

Define the operator $\Delta_L^{(2)}$ in $\mathfrak{L}_2(H, \mu)$ with the everywhere dense domain of definition $D_{\Delta_L^{(2)}}$ putting

$$\Delta_L^{(2)}U = \Delta_L U, \quad D_{\Delta_L^{(2)}} = \mathfrak{P}.$$

Theorem 2.3 *The operator $\Delta_L^{(2)}$ is closable.*

Proof. Let $U_l \in D_{\Delta_L^{(2)}}$, $U_l \to 0$ and $\Delta_L^{(2)} \to G \in \mathfrak{L}_2(H, \mu)$ as $l \to \infty$.

Since $U_l(x) = A_0(l)\mathcal{P}_0 + \sum_{m,q=1}^{N(l)} A_{mq}(l)\mathcal{P}_{mq}(x)$, we have $\lim_{l\to\infty}\{A_0^2(l)\mathcal{P}_0$

$+ \sum_{m,q=1}^{N(l)} A_{mq}^2(l)\} = 0$. Hence $A_{mq} = \lim_{l\to\infty} A_{mq}(l) = 0$ uniformly in m, q.

Since by Theorem 2.2,

$$\Delta_L^2 U_l(x) = 2\Big\{\Big(\frac{1}{\sqrt{2}}\sum_{q=1}^{N(l)} \mu_{2,q} A_{2,q}(l)\Big)\mathcal{P}_0$$

$$+ \sum_{q=1}^{N(l)}\sum_{m=1}^{N(l)-2} \mu_{m+2,q}[(m+2)(m+1)]^{-1/2} A_{m+2,q}(l)\mathcal{P}_{mq}(x)\Big\},$$

we have

$$G_0 = (G, \mathcal{P}_0)_{\mathfrak{L}_2(H,\mu)} = \lim_{l\to\infty}\Big[\sqrt{2}\sum_{q=1}^{N(l)} \mu_{2,q} A_{2,q}(l)\Big] = 0,$$

$$G_{mq} = (G, \mathcal{P}_{mq})_{\mathfrak{L}_2(H,\mu)} = \lim_{l\to\infty} (\Delta_L^{(2)}U_l, \mathcal{P}_{mq})_{\mathfrak{L}_2(H,\mu)}$$

$$= \lim_{l\to\infty} 2\mu_{m+2,q}[(m+2)(m+1)]^{-1/2} A_{m+2,q}(l) = 0,$$

which yields $G = 0$. □

By Theorem 2.2 we compute $\Delta_L U(x)$ for $U \in \mathfrak{P}$, and show that

$$D_{\bar{\Delta}_L^{(2)}} = \Big\{U \in \mathfrak{L}_2(H, \mu) : \Big(\frac{1}{\sqrt{2}}\sum_{q=1}^{\infty} \mu_{2,q} U_{2,q}\Big)^2$$

$$+ \sum_{m,q=1}^{\infty} \mu_{m+2,q}^2[(m+2)(m+1)]^{-1} U_{m+2,q}^2 < \infty\Big\}.$$

2.3 Differential operators of arbitrary order generated by the Lévy Laplacian

The Lévy–Laplace differential expression generates not only operators of the second order, but also operators of any order depending on the choice of the domain of definition of the operator.

The Lévy Laplacian of $F(x)$ is the limit of a sequence of arithmetic means

$$\frac{1}{n}\sum_{k=1}^{n}\xi_k(x),$$

where $\xi_k(x) = (F''(x)f_k, f_k)_H$.

Any function $\xi_k(x)$ on H which is measurable relative to a σ-algebra $\mathfrak{A}$ is a random variable on the probability space $\{H, \mathfrak{A}, \mu\}$. Let $\mathrm{E}\,\xi_k = \int_H \xi_k(x)\mu(dx)$, denote the mathematical expectation and $\mathrm{Var}\,\xi_k = \|\xi_k - \mathrm{E}\,\xi_k\|^2_{\mathcal{L}_2(H,\mu)}$, denote the variance of the random variable ξ_k. The convergence with probability 1 of a sequence of random variables corresponds to the convergence almost everywhere on H in measure μ of a sequence of measurable functions. If, in addition, the sequence $\xi_k(x)$ satisfies the conditions which follow from the strong law of large numbers, and $\lim_{n\to\infty}\frac{1}{n}\sum_{k=1}^n \mathrm{E}\,\xi_k = c$, then for almost all $x \in H$ we have

$$\lim_{n\to\infty}\frac{1}{n}\sum_{k=1}^{n}\xi_k(x) = c.$$

Therefore, it is possible to construct the forms $F(x)$ of degree 2γ such that the Lévy Laplacian decreases the degree of these forms by 2γ, and finally to construct new polynomials by replacing the functions $\|x\|^{2l}_H$ in the polynomials $\mathcal{P}_{mq}(x)$ by the forms $\{F(x)\}^l$.

Put $\gamma_{mq} = a^{2\gamma}_{mq}$ (here $a^{(2)}_{mq} \equiv a_{mq}$), where $\{a^{2\gamma}_{mq}\}^{\infty}_{q=1}$ is the complete system in $\otimes H^m_-$ such that $a^{2\gamma}_{mq} \in \otimes H^m$ at $m = 1, \ldots, 2\gamma - 1$, $a^{2\gamma}_{mq} \in \otimes H^m_-$ at $m = 2\gamma, 2\gamma + 1, \ldots$, and $a^{2\gamma}_{1,q} \equiv \tilde{s}_{1,q}, \ldots, a^{2\gamma}_{2\gamma-1,q} \equiv \tilde{s}_{2\gamma-1,q}$. Then

$$a^{2\gamma}_{2\gamma n-j,q} = \sum_{p=1}^{n} \frac{\mu^{2\gamma}_{2\gamma n-j,q}\mu^{2\gamma}_{2\gamma n-j-2\gamma,q}\cdots\mu^{2\gamma}_{2\gamma p-j,q}[\otimes\delta^{n-p}_{2\gamma}\tilde{\otimes}s_{2\gamma p-j,q}]}{(n-p)!}$$

$$(j = 1, 2, \ldots, 2\gamma - 1),$$

$$a^{2\gamma}_{2\gamma,q} = \mu^{2\gamma}_{2\gamma,q}\delta_{2\gamma} + \tilde{s}_{2\gamma,q},$$

$$a^{2\gamma}_{2\gamma n,q} = \sum_{p=1}^{n} \frac{\mu^{2\gamma}_{2\gamma n,q}\mu^{2\gamma}_{2\gamma n-2\gamma,q}\cdots\mu^{2\gamma}_{2\gamma p,q}[\otimes\delta^{n-p+1}_{2\gamma}\tilde{\otimes}s_{2\gamma p-2\gamma,q}]}{(n-p+1)!}$$

$$+ \tilde{s}_{2\gamma n,q} \qquad (n = 2, 3, \ldots), \qquad\qquad (2.5)$$

where $\delta_{2\gamma} = \sum_{v_1,\ldots,v_{2\gamma}=1}^{\infty} \delta_{v_1 v_2}\delta_{v_2 v_3}\cdots\delta_{v_{2\gamma-1}v_{2\gamma}}T^{\frac{\gamma-1}{\gamma}}e_{v_1}\otimes\ldots\otimes T^{\frac{\gamma-1}{\gamma}}e_{v_{2\gamma}}$, $\delta_2 \equiv \delta$, $\delta_{2\gamma} \in \otimes H^{2\gamma}_-$, $\{e_k\}^{\infty}_1$ is a canonical basis in H, $\{s_{mq}\}^{\infty}_{q=1}$ is a complete sequence of elements in $\otimes H^m_-$ such that $s_{mq} \in \otimes H^m_+$, and the numbers $\mu^{2\gamma}_{mq}$ are obtained in the process of orthogonalization of $a^{2\gamma}_{mq}$, $q = 1, 2, \ldots$.

Denote $P_{a^{2\gamma}_{mq}} \equiv \mathcal{P}^{2\gamma}_{mq}$, $\mathcal{P}^{(2)}_{mq} \equiv \mathcal{P}_{mq}$.

According to Lemma 2.1, the system of polynomials $\mathcal{P}_0, \mathcal{P}_{mq}^{2\gamma}$, $m, q = 1, 2, \ldots$, for each γ, $\gamma = 1, 2, \ldots$, makes an orthonormal basis in $\mathfrak{L}_2(H, \mu)$.

Denote by $\mathfrak{P}_{mq}^{2\gamma}$ the set of all possible linear combinations $A_0^{2\gamma}\mathcal{P}_0 + \sum_{m,q=1}^{N} A_{mq}^{2\gamma}$ ($\mathfrak{P}^{(2)} \equiv \mathfrak{P}$).

Theorem 2.4 *The Lévy Laplacian on $\mathfrak{P}^{(2\gamma)}$ exists, is independent of the choice of basis, and is an operator of order 2γ. It decreases the polynomial degree $\mathcal{P}_{mq}^{2\gamma}$, $m \geq 2\gamma$, by 2γ:*

$$\Delta_L \mathcal{P}_0 = 0, \quad \Delta_L \mathcal{P}_{1,q}(x) = 0, \quad \ldots, \quad \Delta_L \mathcal{P}_{2\gamma-1,q}^{2\gamma}(x) = 0,$$

$$\Delta_L \mathcal{P}_{mq}^{2\gamma}(x) = 2\gamma [m(m-1)\cdots(m-2\gamma+1)]^{-1/2}$$

$$\times [(2\gamma-1)!!]\mu_{mq}^{2\gamma} \mathcal{P}_{m-2\gamma,q}^{2\gamma}(x) \quad (m = 2\gamma, 2\gamma+1, \ldots)$$

for almost all $x \in H$. In particular, on $\mathfrak{P}^\infty$ the Lévy Laplacian is an infinite order operator $\Delta_L \mathcal{P}_0 = 0$, $\Delta_L \mathcal{P}_{mq}^\infty(x) = 0$, $m = 1, 2, \ldots$.

Proof. According to (2.5), for arbitrary basis $\{f_k\}_1^\infty$ in H we have

$$\Phi_{2\gamma n,q}^{2\gamma}(x) = (2\gamma n!)^{-\frac{1}{2}}(a_{2\gamma n,q}^{2\gamma} \otimes x^{2\gamma n})_{\otimes H^{2\gamma n}} = (2\gamma n!)^{-\frac{1}{2}}$$

$$\times \left[\sum_{p=1}^{n} \frac{\mu_{2\gamma n,q}^{2\gamma} \cdots \mu_{2\gamma p,q}^{2\gamma} [F(x)]^{n-p+1}(\tilde{s}_{2\gamma p-2\gamma,q} \otimes x^{2\gamma p-2\gamma})_{\otimes H^{2\gamma p-2\gamma}}}{(n-p+1)!} \right.$$

$$\left. + (\tilde{s}_{2\gamma n,q} \otimes x^{2\gamma n})_{\otimes H^{2\gamma n}} \right],$$

$$F(x) = \sum_{k=1}^{\infty} \sum_{v_1,\ldots,v_{2\gamma}=1}^{\infty} (T^{\frac{\gamma-1}{\gamma}} e_k, f_{v_1})_H \cdots (T^{\frac{\gamma-1}{\gamma}} e_k, f_{v_{2\gamma}})_H$$

$$\times (x, f_{v_1})_H \cdots (x, f_{v_{2\gamma}})_H = \sum_{k=1}^{\infty} (T^{\frac{\gamma-1}{\gamma}} x, e_k)_H^{2\gamma} = \sum_{k=1}^{\infty} \lambda_k^{2\gamma-2}(x, e_k)_H^{2\gamma}.$$

The forms $\Phi_{2\gamma n,q}^{2\gamma} \in \mathfrak{L}_2(H, \mu)$, because $\int_H F^{2l}(x)\mu(dx) < \infty$. One can easily compute integrals $\int_H F^{2l}(x)\mu(dx)$. For example,

$$\int_H F^2(x)\mu(dx) = \int_H \left[\sum_{k=1}^{\infty} \lambda_k^{4\gamma-4}(x, e_k)_H^{4\gamma} \right.$$

$$\left. + \sum_{j,k=1, j \neq k}^{\infty} \lambda_j^{2\gamma-2}(x, e_j)_H^{2\gamma} \lambda_k^{2\gamma-2}(x, e_k)_H^{2\gamma} \right] \mu(dx)$$

$$= \{(4\gamma-1)!! - [(2\gamma-1)!!]^2\}\mathrm{Tr}\, K^2 + [(2\gamma-1)!!\mathrm{Tr}\, K]^2.$$

For functions $F^l(x)(\tilde{s}_{vq}, \otimes x^v)_{\otimes H_v}$ we have

$$\Delta_L[F^l(x)(s_{vq}, \otimes x^v)_{\otimes H^v} = \Delta_L F^l(x) \cdot (s_{vq}, \otimes x^v)_{\otimes H^v}$$
$$+F^l(x) \cdot \Delta_L(s_{vq}, \otimes x^v)_{\otimes H^v} = \Delta_L F^l(x) \cdot (s_{vq}, \otimes x^v)_{\otimes H^v},$$

since $\Delta_L(s_{vq}, \otimes x^v)_{\otimes H^v} = 0$.

Furthermore, $\Delta_L F^l(x) = l F^{l-1}(x) \Delta_L F(x)$.

Let us compute the value $\Delta_L F(x)$. For the above function $F(x)$ we have

$$(F''(x)h, h)_H = 2\gamma(\gamma - 1) \sum_{i=1}^{\infty} (Tx, e_i)_H^{2\gamma-2} (h, e_i)_H^2.$$

Let $f_i = e_i$. Then by (1.2)

$$\Delta_L F(x) = 2\gamma(\gamma - 1) \lim_{n \to \infty} \frac{1}{n} \sum_{k=1}^{n} (Tx, e_k)_H^{2\gamma-2},$$

$\xi_k(x) = (Tx, e_k)_H^{2\gamma-2}$ is a sequence of independent random variables,

$$\mathrm{E}\,\xi_k = \int_H (Tx, e_k)_H^{2\gamma-2} \mu(dx) = \frac{1}{\sqrt{2\pi}} \int_{-\infty}^{\infty} y^{2\gamma-2} e^{-\frac{y^2}{2}} dy = (2\gamma - 3)!!,$$

$$\mathrm{Var}\,\xi_k = \mathrm{E}\,\xi_k^2 - [\mathrm{E}\,\xi_k]^2, \mathrm{E}\,\xi_k^2 = \int_{-\infty}^{\infty} (Tx, e_k)_H^{4\gamma-4} \mu(dx) = (4\gamma - 5)!!,$$

and, therefore, $\sum_{k=1}^{\infty} (\mathrm{Var}\,\xi_k/k^2) < \infty$. According to the strong law of large numbers for independent random variables, with probability 1 $\lim_{n \to \infty} \frac{1}{n} \sum_{k=1}^{n} (\xi_k - \mathrm{E}\,\xi_k) = 0$. Since $\lim_{n \to \infty} \frac{1}{n} \sum_{k=1}^{n} \mathrm{E}\,\xi_k = (2\gamma - 3)!!$, then

$$\lim_{n \to \infty} \frac{1}{n} \sum_{k=1}^{n} \xi_k = (2\gamma - 3)!!$$

almost surely, i.e.,

$$\Delta_L F(x) = 2\gamma(2\gamma - 1)(2\gamma - 3)!!$$

for almost all $x \in H$.

For example, for $\gamma = 2$,

$$\lim_{n \to \infty} \frac{1}{n} \sum_{k=1}^{n} (Tx, e_i)_H^2 = 1,$$

and for $\gamma = 3$,

$$\lim_{n \to \infty} \frac{1}{n} \sum_{k=1}^{n} (Tx, e_i)_H^4 = 3$$

almost surely.

Let $\{f_k\}_1^\infty$ be an arbitrary basis in H. Then by (1.2)

$$\Delta_L F(x) = 2\gamma(2\gamma - 1) \lim_{n\to\infty} \frac{1}{n} \sum_{k=1}^{n} \sum_{i=1}^{\infty} (Tx, e_i)_H^{2\gamma-2}(f_k, e_i)_H^2.$$

Let us represent $s_n(x) = \frac{1}{n}\sum_{k=1}^{n}\sum_{i=1}^{\infty}(Tx, e_i)_H^{2\gamma-2}(f_k, e_i)_H^2$ in the form

$$s_n(x) = \frac{1}{n}\sum_{i=1}^{n}(Tx, e_i)^{2\gamma-2} - \frac{1}{n}\sum_{i=1}^{n}(Tx, e_i)_H^{2\gamma-2}\sum_{k=n+1}^{\infty}(f_k, e_i)_H^2$$

$$+ \frac{1}{n}\sum_{k=1}^{n}\sum_{i=n+1}^{\infty}(Tx, e_i)_H^{2\gamma-2}(f_k, e_i)_H^2 = \sigma_n(x) - \hat{\sigma}_n + \breve{\sigma}_n(x).$$

Since

$$\zeta_i = (Tx, e_i)_H^{2\gamma-2} \quad \text{and} \quad \hat{\zeta}_i = (Tx, e_i)_H^{2\gamma-2}\sum_{k=n+1}^{\infty}(f_k, e_i)_H^2$$

are sequences of independent random variables,

$$\mathrm{E}\,\zeta_i = (2\gamma - 3)!!, \quad \mathrm{E}\,\hat{\zeta}_i = (2\gamma - 3)!!\sum_{k=n+1}^{\infty}(f_k, e_i)^2 \to 0 \quad \text{as} \quad i \to \infty$$

(because $(f_k, e_i)_H^2 \to 0$ as $i \to \infty$, $\sum_{k=n+1}^{\infty}(f_k, e_i)_H^2 \to 0$ as $n \to \infty$) then by the strong law of large numbers $\lim_{n\to\infty}\sigma_n(x) = (2\gamma - 3)!!$, while $\lim_{n\to\infty}\hat{\sigma}_n(x) = 0$ almost surely.

Since

$$\breve{\zeta}_k = \sum_{i=n+1}^{\infty}(Tx, e_i)_H^{2\gamma-2}(f_k, e_i)_H^2$$

is a sequence of arbitrary random variables,

$$\mathrm{E}\,\breve{\zeta}_k = (2\gamma - 3)!!\sum_{i=n+1}^{\infty}(f_k, e_i)_H^2 \to 0 \quad \text{as} \quad k \to \infty,$$

$$\mathrm{E}|\breve{\zeta}_k - \mathrm{E}\,\breve{\zeta}_k| \le c\sum_{i=n+1}^{\infty}(f_k, e_i)^2 \quad \text{and} \quad \sum_{k=1}^{\infty}\frac{1}{k}\mathrm{E}|\breve{\zeta}_k - \mathrm{E}\,\breve{\zeta}_k| < \infty,$$

then by the strong law of large numbers for arbitrary random variables we have $\lim_{n\to\infty}\breve{\sigma}_n(x) = 0$ almost surely.

Thus, almost surely

$$\lim_{n\to\infty} s_n(x) = (2\gamma - 3)!!;$$

that is,

$$\Delta_L F(x) = 2\gamma(2\gamma - 1)[(2\gamma - 3)!!]$$

for almost all $x \in H$.

Since

$$\Delta_L F^l(x) = lF^{l-1}(x)\Delta_L F(x) = \gamma(2\gamma - 1)[(2\gamma - 3)!!]lF^{l-1}(x),$$

the Lévy Laplacian decreases the degree of the form $(\otimes\delta_{2\gamma}^l, \otimes x^{2\gamma l})_{\otimes H^{2\gamma l}}$ by 2γ.

Thus,

$$\Delta_L[F^l(x)(\tilde{s}_{vq}, \otimes x^v)_{\otimes H^v}] = 2\gamma(2\gamma - 1)[(2\gamma - 3)!!]lF^{l-1}(x)(\tilde{s}_{vq}, \otimes x^v)_{\otimes H^v}$$

for almost all $x \in H$. Hence

$$\Delta_L \Phi_{2\gamma n,q}^{2\gamma}(x) = 2\gamma(2\gamma - 1)[2\gamma n(2\gamma n - 1)\ldots(2\gamma n - 2\gamma + 1)]^{-1/2}$$

$$\times [(2\gamma - 3)!!]\mu_{2\gamma n,q}^{2\gamma} \Phi_{2\gamma n-2\gamma,q}^{2\gamma}(x)$$

for almost all $x \in H$.

If $m = 2\gamma(n - 1) + 1, \ldots, m = 2\gamma n - 1$, then similarly we have that almost everywhere on H

$$\Delta_L \Phi_{2\gamma(n-1)+1,q}^{2\gamma}(x) = 2\gamma(2\gamma - 1)[(2\gamma(n - 1) + 1)(2\gamma(n - 1)) \cdots$$

$$\times (2\gamma n - 4\gamma + 2)]^{-1/2}[(2\gamma - 3)!!]\mu_{2\gamma(n-1)+1,q}^{2\gamma} \Phi_{2\gamma n-4\gamma+1,q}^{2\gamma}(x), \ldots,$$

$$\Delta_L \Phi_{2\gamma n-1,q}^{2\gamma}(x) = 2\gamma(2\gamma - 1)[(2\gamma n - 1)(2\gamma n - 2) \cdots (2\gamma n - 2\gamma)]^{-1/2}$$

$$\times [(2\gamma - 3)!!]\mu_{2\gamma n-1,q}^{2\gamma} \Phi_{2\gamma n-2\gamma-1,q}^{2\gamma}(x).$$

This yields

$$\Delta_L \mathcal{P}_{mq}^{2\gamma} = 2\gamma[m(m - 1)\cdots(m - 2\gamma + 1)]^{-1/2}[(2\gamma - 1)!!]\mu_{mq}^{2\gamma} \mathcal{P}_{m-2\gamma,q}^{2\gamma}(x)$$

for almost all $x \in H$. $\square$

Define the operators $\Delta_L^{(2\gamma)}$, $\gamma = 2, 3, \ldots$, in $\mathfrak{L}_2(H, \mu)$ with everywhere dense domains of definition $D_{\Delta_L^{(2\gamma)}}$ putting $\Delta_L^{(2\gamma)}U = \Delta_L U$, $D_{\Delta_L^{(2\gamma)}} = \mathfrak{P}^{2\gamma}$.

Theorem 2.5 *Operator $\Delta_L^{(2\gamma)}$ is closable.*

Proof. Similar to that of Theorem 2.3. $\square$

Applying Theorem 2.4, one can compute $\Delta_L^{(2\gamma)}$ for $U \in \mathfrak{P}^{2\gamma}$, and we show that the family of natural domains of definition of the operators $\bar{\Delta}_L^{(2\gamma)}$ has the form

$$
D_{\bar{\Delta}_L^{(2\gamma)}} = \left\{ U \in \mathfrak{L}_2(H, \mu) : \left[(2\gamma!)^{-1/2} \sum_{q=1}^{\infty} \mu_{2\gamma,q}^{2\gamma} U_{2\gamma,q}^{2\gamma} \right]^2 \right.
$$

$$
\left. + \sum_{m,q=1}^{\infty} \frac{\left[\mu_{m+2\gamma,q}^{2\gamma} U_{m+2\gamma,q}^{2\gamma} \right]^2}{(m + 2\gamma)(m + 2\gamma - 1) \cdots (m + 1)} < \infty \right\}.
$$

3

Symmetric Lévy–Laplace operator

In an infinite-dimensional space, there is no standard measure similar to the Lebesgue measure. If one considers a space with some other measure defined on it, for example, Gaussian measure, even the finite-dimensional Laplacian is not symmetric. Hence one has to symmetrize it.

If $F(x)$ is a cylindrical twice differentiable function, then its Hessian $F''(x)$ is an m-dimensional operator, its gradient $F'(x) \in \mathbb{R}^m$, and therefore the symmetrized Lévy Laplacian on such functions is equal to zero.

An unexpected effect is that the symmetrized Lévy Laplacian can be equal to zero on functions for which the Lévy Laplacian itself is not equal to zero. For example, the symmetrized Lévy Laplacian is equal to zero on functions from the Shilov class. Moreover, it is equal to zero on functions from the natural domain of definition of the Lévy–Laplace operator. Nevertheless it is possible to construct the complete system of orthogonal polynomials such that the symmetrized Lévy Laplacian acts on such polynomials in a non-trivial way. In this case it appears that on these polynomials the Lévy Laplacian has no meaning.

In the final section of this chapter, we again return to the Lévy Laplacian, and we see that it is not always necessary to symmetrize it. Given the Lévy Laplacian (not symmetrized), one can construct symmetric, and even self-adjoint, operators in the Hilbert space $\mathfrak{L}_2(H, \mu)$.

3.1 The symmetrized Lévy Laplacian on functions from the domain of definition of the Lévy–Laplace operator

The symmetrized Lévy Laplacian, if it exists, can be determined by

$$\Delta_C F(x) = \lim_{n \to \infty} \frac{1}{n} \sum_{k=1}^{n} \Big[(F''(x) f_k, f_k)_H - (F'(x), f_k)_H (x, f_k)_{H_+} \Big], \quad (3.1)$$

where $F''(x)$ is the Hessian, $F'(x)$ is the gradient of the function $F(x)$ at the point x and $\{f_k\}_1^\infty$ is the chosen orthonormal basis in H, $f_k \in H_{+2}$. If, in addition, the Lévy Laplacian also exists then

$$\Delta_C F(x) = \Delta_L F(x) - \lim_{n \to \infty} \frac{1}{n} \sum_{k=1}^{n} (F'(x), f_k)_H (x, f_k)_{H_+}.$$

Note that one can rewrite (3.1) in the form

$$\Delta_C F(x) = \lim_{n \to \infty} \frac{1}{n} \sum_{k=1}^{n} \left[d^2 F(x; f_k, f_k) - dF(x; f_k)(x, f_k)_{H_+} \right].$$

If we use Theorem 2.1 for $n = 1$:

$$\int_H dF(x; h)\mu(dx) = \int_H F(x)(h, x)_{H_+}\mu(dx),$$

then, assuming that $U(x)$, $V(x)$ satisfy the conditions of this theorem, we have

$$\int_H V(x)dU(x; h)\mu(dx) = \int_H V(x)U(x)(h, x)_{H_+}\mu(dx)$$

$$- \int_H U(x)dV(x; h)\mu(dx).$$

Applying this formula twice, we obtain

$$\int_H V(x)d^2 U(x; f_k, f_k)\mu(dx) - \int_H V(x)dU(x; f_k)(x, f_k)_{H_+}\mu(dx)$$

$$= - \int_H dV(x; f_k)(x, f_k)_{H_+} U(x)\mu(dx) + \int_H d^2 V(x; f_k; f_k)U(x)\mu(dx).$$

If, in addition, one can pass to the limit under the integral sign, then

$$\int_H V(x) \cdot \Delta_C U(x)\mu(dx) = \int_H \Delta_C V(x) \cdot U(x)\mu(dx).$$

For the Shilov class of functions, $F(x) = \phi(x, ||x||_H^2)$,

$$\Delta_C F(x) = 2 \frac{\partial \phi}{\partial \xi}\Big|_{\xi=||x||_H^2} - 2 \frac{\partial \phi}{\partial \xi}\Big|_{\xi=||x||_H^2} \lim_{n \to \infty} \frac{1}{n} \sum_{k=1}^{n} (x, f_k)_H (x, f_k)_{H_+} = 0$$

for almost all $x \in H$, because almost everywhere on H we have $\lim_{n \to \infty} \frac{1}{n} \sum_{k=1}^{n} (x, f_k)_H (x, f_k)_{H_+} = 1$.

Let us show that on polynomials $\mathcal{P}_{mq}(x)$, $m, q = 1, 2, \ldots$, from the domain of definition of the Lévy–Laplace operator, the symmetrized Lévy Laplacian is equal to zero for almost all $x \in H$.

The polynomials $\mathcal{P}_{mq}(x)$ consist of the forms $\Phi_{a_{mq}}(x)$, and they are from

$$H_{lv}(x) = \frac{1}{l!} ||x||_H^{2l} (\tilde{s}_{\gamma q}, \otimes x)_{\otimes H^v};$$

see Section 2.2.

For functions $H_{lv}(x)$ we have

$$\Delta_L H_{lv}(x) = \frac{2}{(l-1)!} ||x||_H^{2(l-1)} (\tilde{s}_{vq}, \otimes x^v)_{\otimes H^v}.$$

Since

$$(H'_{lv}(x), h)_H = \frac{2}{(l-1)!} ||x||_H^{2(l-1)} (x, h)_H (\tilde{s}_{vq}, \otimes x^v)_{\otimes H^v}$$
$$+ \frac{1}{l!} ||x||_H^{2l} v (\tilde{s}_{vq}, \otimes x^{v-1} \otimes h)_{\otimes H^v},$$

then, taking into account

$$\lim_{n \to \infty} \frac{1}{n} \sum_{k=1}^{n} (\tilde{s}_{vq}, \otimes x^{v-1} \otimes f_k)_{\otimes H^v} (x, f_k)_{H_+} = 0,$$

since $\tilde{s}_{vq} \in \otimes H_+^v$, we obtain

$$\Delta_C H_{lv}(x)$$
$$= \frac{2}{(l-1)!} ||x||_H^{2(l-1)} \lim_{n \to \infty} \frac{1}{n} \sum_{k=1}^{n} \Big[1 - (x, f_k)_H (x, f_k)_{H_+} \Big] (\tilde{s}_{vq}, \otimes x^v)_{\otimes H^v}.$$

We now show that

$$s_n(x) = \frac{1}{n} \sum_{k=1}^{n} \Big[1 - (x, f_k)_H (x, f_k)_{H_+} \Big]$$

converges to zero almost everywhere on H.

Let $f_k = e_k$. Then

$$s_n(x) = \frac{1}{n} \sum_{k=1}^{n} \Big[1 - (Tx, e_k)_H^2 \Big],$$

$\zeta_k = 1 - (Tx, e_k)_H^2$ is a sequence of independent random variables,

$$E\zeta_k = 1 - \frac{1}{\sqrt{2\pi}} \int_{-\infty}^{\infty} y^2 e^{-\frac{1}{2}y^2} dy = 0,$$

$$Var\zeta_k = \frac{1}{\sqrt{2\pi}} \int_{-\infty}^{\infty} (1 - y^2)^2 e^{-\frac{1}{2}y^2} dy = 2 \quad \text{and} \quad \sum_{k=1}^{\infty} \frac{Var\zeta_k}{k^2} < \infty.$$

By the strong law of large numbers for independent random variables we deduce that $\lim_{n \to \infty} s_n(x) = 0$ with probability equal to unity. This means that

$$\Delta_C \left[\frac{1}{l!} ||x||_H^{2l} \right] = 0$$

for almost all $x \in H$.

Let $\{f_k\}_1^{\infty}$ be an arbitrary basis in H, $f_k \in H_{+2}$. Then

$$s_n = 1 - \frac{1}{n} \sum_{k=1}^{n} \sum_{i=1}^{\infty} (x, f_k)_H (T^2 x, e_i)_H (f_k, e_i)_H$$

$$= \left[1 - \frac{1}{n} \sum_{i=1}^{n} (T^2 x, e_i)_H (x, e_i)_H \right] - \frac{1}{n} \sum_{k=1}^{n} (x, f_k)_H \sum_{i=n+1}^{\infty} (T^2 x, e_i)_H (f_k, e_i)_H$$

$$+ \frac{1}{n} \sum_{i=1}^{n} (T^2 x, e_i)_H \sum_{k=n+1}^{\infty} (x, f_k)_H (f_k, e_i)_H = \sigma_n(x) - \hat{\sigma}_n(x) + \check{\sigma}_n(x).$$

Since $\zeta_i = [1 - (T^2 x, e_i)_H (x, e_i)_H]$ is a sequence of independent random variables, then (as shown above) $\lim_{n \to \infty} \sigma_n = 0$ almost surely.

Since

$$\hat{\zeta}_k = (x, f_k)_H \sum_{i=n+1}^{\infty} (T^2 x, e_i)_H (f_k, e_i)_H$$

is a sequence of arbitrary random variables,

$$E\hat{\zeta}_k = \sum_{k=n+1}^{\infty} (f_k, e_i)_H^2 \to 0 \quad \text{as} \quad k \to \infty$$

(because $(f_k, e_i)_H \to 0$ as $k \to \infty$, $\sum_{i=n+1}^{\infty} (f_k, e_i)_H^2 \to 0$ as $n \to \infty$),

$$E|\hat{\zeta}_k - E\hat{\zeta}_k| \le c \sum_{i=n+1}^{\infty} (f_k, e_i)_H^2 \quad \text{and} \quad \sum_{k=1}^{\infty} \frac{1}{k} E|\hat{\zeta}_k - E\hat{\zeta}_k| < \infty,$$

then by the strong law of large numbers for arbitrary random variables we have $\lim_{n \to \infty} \hat{\sigma}_n = 0$ almost surely. The law of large numbers also yields that $\lim_{n \to \infty} \check{\sigma}_n = 0$ almost surely.

Therefore, almost surely $\lim_{n\to\infty} s_n = 0$, which means that $\Delta_C[\frac{1}{l}||x||_H^{2l}] = 0$ for almost all $x \in H$.

Thus,

$$\Delta_C H_{l\nu}(x) = 0$$

for almost all $x \in H$. Hence, almost everywhere on H we have

$$\Delta_C \Phi_{a_{2n,q}}(x) = 0.$$

If $m = 2n + 1$, then we similarly have that almost everywhere on H

$$\Delta_C \Phi_{a_{2n+1,q}}(x) = 0.$$

Therefore

$$\Delta_C \mathcal{P}_{mq}(x) = 0$$

for almost all $x \in H$, $m, q = 1, 2, \ldots$.

3.2 The Lévy Laplacian on functions from the domain of definition of the symmetrized Lévy–Laplace operator

In Section 3.1 we showed that the symmetrized Lévy Laplacian is equal to zero on a wide class of functions, in particular, on polynomials from the natural domain of definition of the Lévy–Laplace operator.

Do functions from $\mathfrak{L}_2(H, \mu)$ exist for which the symmetrized Lévy Laplacian is not equal to zero? The symmetrized Lévy Laplacian of $G(x)$ is the limit of the sequence of arithmetic means

$$\frac{1}{n} \sum_{k=1}^{n} \zeta_k(x), \quad \zeta_k(x) = (G''(x)f_k, f_k)_H - (G'(x), f_k)_H(x, f_k)_{H_+}.$$

Even when $\int_H \zeta_k(x)\mu(dx) = 0$ one would expect that $\Delta_C G(x) \neq 0$, if $G(x)$ is such a function that the sequence ζ_k does not obey the law of large numbers. If in addition $G(x) \in \mathfrak{L}_2(H, \mu)$, then one can construct new polynomials by replacing the functions $||x||_H^{2l}$ in the $\mathcal{P}_{mq}(x)$ by the $G(x)^l$.

We construct in $\mathfrak{L}_2(H, \mu)$ a complete orthonormal system of polynomials $Q_0, Q_{mq}(x)$, $m, q = 1, 2, \ldots$, such that $\Delta_C Q_{mq}(x) \neq 0$ and it belongs to $\mathfrak{L}_2(H, \mu)$.

Assume that the correlation operator K of the measure μ is such that $K^{1/2}\chi^{1/2}(K^{-1/2})$ is a Hilbert–Schmidt operator, where $\xi = \chi^2(\eta)$ is the inverse function of $\eta = \lambda(\xi)$, and $\lambda(\xi)$ is a certain monotonically increasing function that is continuous for $\xi > 0$ and such that $\lambda(k) = \lambda_k$; $\chi(\lambda_k) = \sqrt{k}$.

We put $\gamma_{mq} = \alpha_{mq}$, where $\{\alpha_{mq}\}_{q=1}^{\infty}$ is a complete orthonormal system of elements in $\otimes H_{-}^{m}$ such that

$$\alpha_{2n,q} = \sum_{p=1}^{n} \frac{v_{2n,q} v_{2n-2,q} \cdots v_{2p,q} \left[\otimes \sigma^{n-p+1} \tilde{\otimes} s_{2p-2,q}\right]}{(n-p+1)!} + \tilde{s}_{2n,q},$$

$$\alpha_{2n-1,q} = \sum_{p=1}^{n} \frac{v_{2n-1,q} v_{2n-3,q} \cdots v_{2p-1,q} \left[\otimes \sigma^{n-p} \tilde{\otimes} s_{2p-1,q}\right]}{(n-p)!}$$

$$(n = 1, 2, \ldots), \tag{3.2}$$

where

$$\sigma = \sum_{j,k=1}^{\infty} (\chi(T)g_j, g_k)_H g_j \otimes g_k,$$

$\{g_k\}_1^{\infty}$ is a fixed orthonormal basis in H, $g_k \in H_{+2}$, $\{s_{mq}\}_{q=1}^{\infty}$ is a complete sequence of elements in $\otimes H_{-}^{m}$ such that $s_{mq} \in \otimes H_{+}^{m}$, $m = 1, 2, \ldots$, and the numbers v_{mq} are obtained by the orthogonalization of α_{mq}, $m = 1$, $2, \ldots$.

Since by assumption, $T^{-1}\chi^{1/2}(T)$ is a Hilbert–Schmidt operator, the forms $\Phi_{\alpha_{mq}}(x) \in \mathcal{L}_2(H, \mu)$.

Denote $P_{\alpha_{mq}} \equiv Q_{mq}$.

By Lemma 1.2 the system of polynomials Q_0, Q_{mq}, $m = 1, 2, \ldots$, forms an orthonormal basis in $\mathcal{L}_2(H, \mu)$.

Theorem 3.1 $\Delta_C Q_{mq}(x) \neq 0$, $\Delta_C Q_{mq}(x) \in \mathcal{L}_2(H, \mu)$.

Proof. By (3.2), for an arbitrary basis $\{f_k\}_1^{\infty}$ in H, $f_k \in H_{+2}$, we have

$$\Phi_{\alpha_{2n,q}}(x) = (2n!)^{-\frac{1}{2}} (\alpha_{2n,q}, \otimes x^{2n})_{\otimes H^{2n}}$$

$$= (2n!)^{-\frac{1}{2}} \left[\sum_{p=1}^{n} \frac{v_{2n,q} \cdots v_{2p,q}\{G(x)\}^{n-p+1}(\tilde{s}_{2p-2,q}, \otimes x^{2p-2})_{\otimes H^{2p-2}}}{(n-p+1)!}\right.$$

$$\left. + (\tilde{s}_{2n,q}, \otimes x^{2n})_{\otimes H^{2n}}\right],$$

$$G(x) = \sum_{j,i=1}^{\infty} (\chi(T)f_j, f_i)_H (x, f_j)_H (x, f_i)_H.$$

Let $f_i = e_i$. Then

$$G(x) = \sum_{k=1}^{\infty} \sqrt{k}(x, e_k)_H^2$$

and, by (3.1),

$$\Delta_C G(x) = 2 \lim_{n \to \infty} \frac{1}{n} \sum_{k=1}^{n} \sqrt{k}[1 - \lambda_k^2(x, e_k)_H^2].$$

We show that the sequence $p_n(x) = \frac{1}{n} \sum_{k=1}^{n} \sqrt{k} \, [1 - \lambda_k^2(x, e_k)^2]$ converges in $\mathcal{L}_2(H, \mu)$. For $m < n$ we have

$$\int_H [p_n(x) - p_m(x)]^2 \mu(dx) = \frac{2}{n^2} \sum_{k=1}^{n} k - \frac{4}{nm} \sum_{k=1}^{m} k + \frac{2}{m^2} \sum_{k=1}^{m} k \to 0$$

as $m, n \to \infty$. Now $\int_H p_n^2(x)\mu(dx) = \frac{2}{n^2} \sum_{k=1}^{n} k = \frac{n+1}{n}$, and

$$\|\Delta_C G(x)\|_{\mathcal{L}_2(H,\mu)} = 2.$$

Let $\{f_k\}_1^\infty$ be an arbitrary basis in H, $f_k \in H_{+2}$. Then

$$(G'(x), h)_H = 2 \sum_{i,j=1}^{\infty} (\chi(T)f_i, f_j)_H(x, f_j)_H(h, f_i)_H,$$

$$(G''(x)h, h)_H = 2 \sum_{i,j=1}^{\infty} (\chi(T)f_i, f_j)_H(h, f_j)_H(h, f_i)_H,$$

and by (3.1),

$$\Delta_C G(x) = 2 \lim_{n \to \infty} \frac{1}{n} \sum_{k=1}^{n} [(\chi(T)f_k, f_k)_H - (\chi(T)x, f_k)_H(T^2x, f_k)_H].$$

Similarly to the previous case, we can easily show that

$$q_n(x) = \frac{1}{n} \sum_{k=1}^{n} [(\chi(T)f_k, f_k)_H - (\chi(T)x, f_k)_H(T^2x, f_k)_H]$$

is a fundamental sequence. Taking into account that

$$\int_H (x, \alpha)_H(x, \beta)_H \mu(dx) = (\alpha, \beta)_{H_-},$$

$$\int_H (x, \alpha)_H(x, \beta)_H(x, \gamma)_H(x, \sigma)_H \mu(dx) = (\alpha, \beta)_{H_-}(\gamma, \sigma)_{H_-}$$

$$+ (\alpha, \gamma)_{H_-}(\beta, \sigma)_{H_-} + (\alpha, \sigma)_{H_-}(\beta, \gamma)_{H_-}, \quad \alpha, \beta, \gamma, \sigma \in H_-,$$

we obtain that

$$\int_H q_n^2(x)\mu(dx)$$

$$= \frac{1}{n^2} \int_H \left\{ \sum_{j,k=1, j\neq k}^{n} \left[(\chi(T)f_k, f_k)_H - (\chi(T)x, f_k)_H (T^2 x, f_k)_H \right] \right.$$

$$\times \left[(\chi(T)f_j, f_j)_H - (\chi(T)x, f_j)_H (T^2 x, f_j)_H \right]$$

$$\left. + \sum_{k=1}^{n} \left[(\chi(T)f_k, f_k)_H - (\chi(T)x, f_k)_H (T^2 x, f_k)_H \right]^2 \right\} \mu(dx)$$

$$= \frac{1}{n^2} \sum_{j,k=1}^{n} \left[(\chi(T)f_k, f_j)_H^2 + (\chi(T)f_k, \chi(T)f_j)_{H_-}(f_k, f_j)_{H_+} \right],$$

and $\lim\limits_{n\to\infty} \int_H q_n^2(x)\mu(dx) = 1$. Therefore,

$$\||\Delta_C G(x)|\|_{\mathcal{L}_2(H,\mu)} = 2.$$

The polynomials $Q_{mq}(x)$ consist of the forms $\Phi_{\alpha_{mq}}(x)$, and they are from

$$Z_{lv}(x) = \frac{1}{l!} G^l(x)(\tilde{s}_{vq}, \otimes x^v)_{\otimes H^v}.$$

For functions Z_{lv} we have

$$\Delta_C Z_{lv}(x) = \frac{1}{(l-1)!} G^{l-1}(x)\Delta_C G(x)(\tilde{s}_{vq}, \otimes x^v)_{\otimes H^v}$$

(since $\tilde{s}_{vq} \in \otimes H_+^v$). Therefore, $\Delta_C \Phi_{\alpha_{2n,q}}(x) \in \mathcal{L}_2(H, \mu)$ (since $\Phi_{\alpha_{2n',q'}}(x) \in \mathcal{L}_2(H, \mu)$).

If $m = 2n + 1$, then similarly we have $\Delta_C \Phi_{\alpha_{2n+1,q}}(x) \in \mathcal{L}_2(H, \mu)$. Therefore, $\Delta_C Q_{mq}(x) \in \mathcal{L}_2(H, \mu)$. $\qquad\square$

We show that the Lévy Laplacian has no meaning on the polynomials Q_{mq}, $m = 2, 3, \ldots; q = 1, 2, \ldots$.
If $f_k = e_k$, then $G(x) = \sum_{k=1}^{\infty} \sqrt{k}(x, e_k)_H^2$, and

$$\Delta_L G(x) = 2 \lim_{n\to\infty} \frac{1}{n} \sum_{k=1}^{n} \sqrt{k} = \infty.$$

If $\{f_k\}_1^{\infty}$ is an arbitrary basis in H, $f_k \in H_{+2}$, then

$$G(x) = \sum_{j,k=1}^{\infty} (\chi(T)f_j, f_k)_H (x, f_j)_H (x, f_k)_H,$$

and

$$\Delta_L G(x) = 2 \lim_{n \to \infty} \frac{1}{n} \sum_{k=1}^{n} (\chi(T) f_k, f_k)_H.$$

Since

$$\sum_{k=1}^{n} (\chi(T) f_k, f_k)_H = \sum_{k=1}^{n} (P\chi(T) P f_k, f_k)_H = \sum_{k=1}^{n} (P\chi(T) P \varphi_k, \varphi_k)_H$$

$$= \sum_{k=1}^{n} (\chi(T) e_{n_k}, e_{n_k})_H = \sum_{k=1}^{n} \chi(\lambda_{n_k}),$$

where P is an orthogonal projection on the space with the basis $\{f_k\}_1^n$, φ_k are eigenvectors of operator $P\chi(T)P$, $P\varphi_k = e_{n_k}$, then

$$\Delta_L G(x) = 2 \lim_{n \to \infty} \frac{1}{n} \sum_{k=1}^{n} \chi(\lambda_{n_k}) = \infty,$$

since $\chi(\lambda_{n_k}) \to \infty$ as $k \to \infty$.

3.3 Self-adjointness of the non-symmetrized Lévy–Laplace operator

Let us consider a statement which certainly can not be valid for the finite-dimensional Laplacian. We show that, given a non-symmetrized Lévy Laplacian, one can construct a self-adjoint operator in the space of functions square-integrable in Gaussian measure.

We denote by $\mathfrak{T}$ the set of all functions of the form

$$V(x) = \varphi(Q(x)) S(x),$$

where $Q(x) = \|x\|_H^2$, $S(x)$ are arbitrary twice strongly differentiable harmonic in H functions from $\mathfrak{L}_2(H, \mu)$, $\varphi(\xi)$ is an arbitrary fixed positive function defined and differentiable on $[0, \infty)$ satisfying the Lipschitz condition with the Lipschitz constant c. Let, in addition, $V(x)$ and $\left(\varphi'(\|x\|_H^2)/\varphi(\|x\|_H^2)\right) V(x) \in \mathfrak{L}_2(H, \mu)$.

The set $\mathfrak{T}$ is linear.

To show that it is everywhere dense in $\mathfrak{L}_2(H, \mu)$, it is sufficient to show that the set $\hat{\mathfrak{T}}$ is everywhere dense, where $\hat{\mathfrak{T}}$ is the set of all functions

$$\Phi(x) = \varphi(\|x\|_H^2)\Psi(x),$$

where $\Psi(x) \in \mathfrak{C}$. Here $\mathfrak{C}$ is the set of cylindrical twice strongly differentiable functions. If $\Psi \in \mathfrak{C}$, that is $\Psi(x) = \Psi(Px)$, P is a projection on an

m-dimensional subspace, then its Hessian $\Psi''(x)$ is a finite-dimensional (m-dimensional) operator, and

$$\Delta_L \Psi(x) = \lim_{n \to \infty} \frac{1}{n} \sum_{k=1}^{m} (\Psi''(x) f_k, f_k)_H = 0.$$

Therefore, $\Psi(x)$ is a harmonic function in H, and $\hat{\mathfrak{T}} \subset \mathfrak{T}$.

Let $U \in \mathfrak{L}_2(H, \mu)$. The set $\mathfrak{C}$ is everywhere dense in $\mathfrak{L}_2(H, \mu)$, i.e., for all ε_1 there exists $\Psi_0 \in \mathfrak{C}$ such that

$$\|U - \Psi_0\|_{\mathfrak{L}_2(H,\mu)} \leq \varepsilon_1.$$

Since $\Psi_0 \in \mathfrak{C}$, we have $\Psi_0(x) = \Psi_0(Px)$ for some projection on an m-dimensional subspace with a basis $\{f_k\}_1^m$. Choose

$$\Phi_0(x) = \varphi(\|x\|_H^2) \Psi_0(x)/\varphi(\|Px\|_H^2);$$

$\Phi_0 \in \hat{\mathfrak{T}}$, because $\Psi_0(x)/\varphi(\|Px\|_H^2) \in \mathfrak{C}$. Then

$$\|U - \Phi_0\|_{\mathfrak{L}_2(H,\mu)} \leq \|U - \Psi_0\|_{\mathfrak{L}_2(H,\mu)} + \|\Psi_0 - \Phi_0\|_{\mathfrak{L}_2(H,\mu)}$$
$$\leq \varepsilon_1 + \|\Psi_0 - \Phi_0\|_{\mathfrak{L}_2(H,\mu)}.$$

We have

$$\|\Psi_0 - \Phi_0\|_{\mathfrak{L}_2(H,\mu)}^2 = \int_H [\Psi_0(x) - \varphi(\|x\|_H^2) \Psi_0(x)/\varphi(\|Px\|_H^2)]^2 \mu(dx)$$

$$= \int_H [\Psi_0(x)/\varphi(\|Px\|_H^2)]^2 [\varphi(\|Px\|_H^2) - \varphi(\|x\|_H^2)]^2 \mu(dx)$$

$$\leq c^2 \int_H [\Psi_0(x)/\varphi(\|Px\|_H^2)]^2 \Big[\sum_{k=m+1}^{\infty} (x, f_k)_H^2 \Big]^2 \mu(dx),$$

because the function $\varphi(\xi)$ satisfies the Lipschitz condition.

Since $\Psi_0(x)/\varphi(\|Px\|_H^2)$ depends only on $(x, f_1)_H, \ldots, (x, f_m)_H$, and $\sum_{k=m+1}^{\infty}(x, f_k)_H^2$ depends on $(x, f_{m+1})_H$, $(x, f_{m+2})_H, \ldots$, we have

$$\|\Psi_0 - \Phi_0\|_{\mathfrak{L}_2(H,\mu)}^2$$

$$\leq c^2 \int_H [\Psi_0(x)/\varphi(\|Px\|_H^2)]^2 \mu(dx) \int_H \Big[\sum_{k=m+1}^{\infty} (x, f_k)_H^2 \Big]^2 \mu(dx)$$

$$= c^2 \int_H [\Psi_0(x)/\varphi(\|Px\|_H^2)]^2 \mu(dx)$$

$$\times \Big\{ \Big[\sum_{k=m+1}^{\infty} (Kf_k, f_k)_H \Big]^2 + 2 \sum_{k=m+1}^{\infty} (K^2 f_k, f_k)_H \Big\}.$$

Since $\Psi_0(x)/\varphi(\|Px\|_H^2) \in \mathfrak{L}_2(H, \mu)$, and $\sum_{k=1}^\infty (Kf_k, f_k)_H = \mathrm{Tr}K < \infty$, $\sum_{k=1}^\infty (K^2 f_k, f_k)_H = \mathrm{Tr}K^2 < \infty$ for all $\{f_k\}_1^\infty$ in H (K is a trace class operator), and we have

$$\|\Phi_0 - \Psi_0\|_{\mathfrak{L}_2(H,\mu)} \le \varepsilon_2.$$

Therefore,

$$\|U - \Phi_0\|_{\mathfrak{L}_2(H,\mu)} \le \varepsilon_1 + \varepsilon_2 = \varepsilon$$

and hence $\hat{\hat{\mathfrak{T}}}$ is dense everywhere in $\mathfrak{L}_2(H, \mu)$.

Lemma 3.1 *The Lévy Laplacian on $\mathfrak{T}$ exists, is independent of the choice of the basis, and is an operator of multiplication by the function $2\varphi'(\|x\|_H^2)/\varphi(\|x\|_H^2)$:*

$$\Delta_L V(x) = \frac{2\varphi'(\|x\|_H^2)}{\varphi(\|x\|_H^2)} V(x) \qquad (V \in \mathfrak{T}).$$

Proof. By Corollary 2 to (1.4), we have

$$\Delta_L V(x) = \Delta_L[\varphi(\|x\|_H^2)S(x)] = \Delta_L\varphi(\|x\|_H^2) \cdot S(x) + \varphi(\|x\|_H^2) \cdot \Delta_L S(x).$$

By (1.4) for $m = 1$, we obtain

$$\Delta_L\varphi(\|x\|_H^2) = \varphi'(\|x\|_H^2)\Delta_L\|x\|_H^2 = 2\varphi'(\|x\|_H^2) \quad (\text{since } \Delta_L\|x\|_H^2 = 2).$$

In addition, since $S(x)$ is harmonic, $\Delta_L S(x) = 0$.

Therefore,

$$\Delta_L V(x) = 2\varphi'(\|x\|_H^2)S(x) = \frac{2\varphi'(\|x\|_H^2)}{\varphi(\|x\|_H^2)} V(x).$$

$\square$

Lemma 3.2 *If $\Delta_L V(x) = (2\varphi'(\|x\|_H^2)/\varphi(\|x\|_H^2))V(x)$, then the function $V(x)$ has the form*

$$V(x) = \varphi(\|x\|_H^2)S(x),$$

where $S(x)$ is an arbitrary harmonic function.

Proof. It follows from $\Delta_L V(x) = \big(2\varphi'(\|x\|_H^2)/\varphi(\|x\|_H^2)\big) V(x)$, for $V(x) \ne 0$, that

$$\frac{\Delta_L V(x)}{V(x)} = \frac{\Delta_L\varphi(\|x\|_H^2)}{\varphi(\|x\|_H^2)}.$$

Hence, $\Delta_L\{\ln|V(x)|\} = \Delta_L\{\ln\varphi(\|x\|_H^2)\}$, i.e.,

$$\Delta_L\left\{\ln\frac{|V(x)|}{\varphi(\|x\|_H^2)}\right\} = 0.$$

Therefore

$$\ln\frac{|V(x)|}{\varphi(\|x\|_H^2)} = G(x),$$

where $G(x)$ is an arbitrary harmonic function. Hence,

$$V(x) = \varphi(\|x\|_H^2)e^{G(x)} = \varphi(\|x\|_H^2)S(x),$$

where $S(x)$ is an arbitrary harmonic function. $\qquad\qquad\square$

We define an operator L in $\mathfrak{L}_2(H, \mu)$ with everywhere dense domain of definition D_L, putting

$$LU = \Delta_L U, \qquad D_L = \mathfrak{T}.$$

Theorem 3.2 *The operator L is essentially self-adjoint.*

Proof. The operator L is symmetric, since D_L is dense in $\mathfrak{L}_2(H, \mu)$ and by Lemma 3.1

$$(LV_1, V_2)_{\mathfrak{L}_2(H,\mu)} = \int_H \frac{2\varphi'(\|x\|_H^2)}{\varphi(\|x\|_H^2)} V_1(x)V_2(x)\mu(dx)$$

$$= (V_1, LV_2)_{\mathfrak{L}_2(H,\mu)} \qquad \text{for all} \qquad V_1, V_2 \in D_L.$$

Let us show that L is essentially self-adjoint.

Consider the operator L^* adjoint to L in $\mathfrak{L}_2(H, \mu)$. Let $Z \in D_{L^*}$. Then for any $V \in D_L$ we have

$$(LV, Z)_{\mathfrak{L}_2(H,\mu)} = (V, L^*Z)_{\mathfrak{L}_2(H,\mu)};$$

that is,

$$\int_H \frac{2\varphi'(\|x\|_H^2)}{\varphi(\|x\|_H^2)} V(x)Z(x)\mu(dx) = \int_H V(x)(L^*Z)(x)\mu(dx)$$

or, taking onto account that $V(x) = \varphi(\|x\|_H^2)S_V(x)$, we get

$$\int_H \varphi(\|x\|_H^2)S_V(x)\left[\frac{2\varphi'(\|x\|_H^2)}{\varphi(\|x\|_H^2)} Z(x) - (L^*Z)(x)\right]\mu(dx) = 0.$$

This equality holds for functions having the form $\varphi(\|x\|_H^2)S(x)$ where $S(x) \in \mathfrak{C}$ (the set of cylindrical functions) and $\varphi(\xi) > 0$ (for example, it holds if $S(x)$ belongs to the complete orthonormal Hermite–Fourier polynomial system). Hence almost everywhere on H we have

$$L^*Z = \frac{2\varphi'(\|x\|_H^2)}{\varphi(\|x\|_H^2)} Z.$$

Thus for all $Z \in D_{L^*}$

$$\frac{2\varphi'(\|x\|_H^2)}{\varphi(\|x\|_H^2)} Z \in \mathfrak{L}_2(H, \mu) \quad \text{and} \quad L^*Z = \frac{2\varphi'(\|x\|_H^2)}{\varphi(\|x\|_H^2)} Z.$$

By the equality

$$\int_H \left[\frac{2\varphi'(\|x\|_H^2)\, V(x)}{\varphi(\|x\|_H^2)} \right] Z(x)\mu(dx) = \int_H U(x) \left[\frac{2\varphi'(\|x\|_H^2)\, Z(x)}{\varphi(\|x\|_H^2)} \right] \mu(dx)$$

it is easy to see that for any $Z \in \mathfrak{L}_2(H, \mu)$, $L^*Z = \left(2\varphi'(\|x\|_H^2)/\varphi(\|x\|_H^2) \right) Z$ and $\left(2\varphi'(\|x\|_H^2)/\varphi(\|x\|_H^2) \right) Z \in \mathfrak{L}_2(H, \mu)$.

This shows that

$$D_{L^*} = \left\{ Z \in \mathfrak{L}_2(H, \mu) : \quad \frac{2\varphi'(\|x\|_H^2)}{\varphi(\|x\|_H^2)} Z \in \mathfrak{L}_2(H, \mu) \right\}.$$

The domain D_{L^*} is dense in $\mathfrak{L}_2(H, \mu)$ since $D_{L^*} \supset D_L$. The operator $L^*Z = \left(2\varphi'(\|x\|_H^2)/\varphi(\|x\|_H^2) \right) Z$, $Z \in D_{L^*}$, is a self-adjoint operator.

By the self-adjointness of L^* we deduce that L is essentially self-adjoint.

$\square$

4

Harmonic functions of infinitely many variables

It is well known that the only harmonic function of a single variable is a linear function. It is also known that in a multi-dimensional case the number of harmonic functions grows. This number increases rapidly if we are interested in infinite-dimensional harmonic functions associated with the Lévy Laplacian.

Let us give several examples (which have no finite-dimensional counterparts) of harmonic functions defined on a Hilbert space H:

1. Cylindrical twice differentiable functions.

   Let $F(x)$ be a cylindrical twice differentiable function $F(x) = F(Px)$, and P be a projection on m-dimensional subspace. Then the Hessian $F''(x)$ is a finite-dimensional (m-dimensional) operator and (1.2) leads to

   $$\Delta_L F(x) = \lim_{n \to \infty} \frac{1}{n} \sum_{k=1}^{m} (F''(x) f_k, f_k)_H = 0;$$

   here $\{f_k\}_1^\infty$ is an orthonormal basis in H.

2. Twice differentiable functions $F(x)$, whose Hessians $F''(x)$ are compact operators in H.

   Indeed, since Hessian $F''(x)$ is a compact self-adjoint operator in H, we have $(F''(x) f_k, f_k)_H \to 0$ as $k \to \infty$ and, according to (1.2),

   $$\Delta_L F(x) = \lim_{n \to \infty} \frac{1}{n} \sum_{k=1}^{n} (F''(x) f_k, f_k)_H = 0$$

   for arbitrary orthonormal basis $\{f_k\}_1^\infty$ in H.

   For example, if $H = L_2(a, b)$, and the second differential of a function $F(x)$ has the regular form (Example 1.2), then

   $$F''(x)h(s) = \int_a^b \frac{\delta^2 F(x)}{\delta x(s) \delta x(\tau)} h(s) \, ds$$

$(\delta^2 F(x)/\delta x(s)\delta x(\tau)$ is a continuous function with respect to s, τ, $\delta^2 F(x)/\delta x(s)\delta x(\tau) = \delta^2 F(x)/\delta x(\tau)\delta x(s))$, and $F''(x)$ is a compact operator in $L_2(a, b)$.

Another example of this kind is given by $H = l_2$, and $F(x) = \sum_{k=1}^{\infty} x_k^m$ for integer numbers $m \geq 3$. In this case the Hessian $F''(x)$ is a matrix $m(m - 1)\|\delta_{ik} x_k^{m-2}\|_{i,k=1}^{\infty}$ (here δ_{ik} is the Kronecker symbol), $x_k^{m-2} \to 0$ as $k \to \infty$, and $F''(x)$ is a compact operator in l_2.

3. Functions of the form $F(x) = f(U_1(x), \ldots, U_m(x))$, where $f(u_1, \ldots, u_m)$ is a twice continuously differentiable function of m variables over the set $\{U_1(x), \ldots, U_m(x)\}$ in $\mathbb{R}^m$, and functions $U_j(x)$ are harmonic in the whole space H, $j = 1, \ldots, m$.

Indeed, since $\Delta_L U_j(x) = 0$, $j = 1, \ldots, m$, we derive, using (1.4),

$$\Delta_L F(x) = \sum_{j=1}^{m} \frac{\partial F}{\partial u_j}\Big|_{u_j=U_j(x)} \Delta_L U_j(x) = 0.$$

4. $$H = l^2, \quad F(x) = \sum_{k=1}^{\infty} \alpha_k \lambda_k^{2m-2} x_k^{2m} \quad (\alpha_k = O(1/\ln^{1+\varepsilon} k), \quad \varepsilon > 0),$$

$$F(x) = \sum_{k=1}^{\infty} \prod_{j=1}^{k} \lambda_j x_j \lambda_k^{2m-1} x_k^{2m}, \quad F(x) = \sum_{k=1}^{\infty} \lambda_k^{2m-1} x_k^{2m+1},$$

$$F(x) = \sum_{k=1}^{\infty} \left(\frac{\lambda_k^{2m} x_k^{2m+2}}{(2m - 1)!!(2m + 2)(2m + 1)} - \frac{x_k^2}{2} \right),$$

$m \geq 1$ is an integer, $x_k = (x, e_k)_{l_2}$, $e_k = (\underbrace{0, \ldots, 1}_{k}, 0, \ldots)$, $K^{-1/2} e_k = \lambda_k e_k$, $k = 1, 2, \ldots$.

As a result of the strong law of large numbers, all these functions are harmonic μ–almost everywhere on l_2.

4.1 Arbitrary second-order derivatives

Theorem 4.1 *Let $F(x)$ be a twice differentiable function with respect to a subspace H_α for some $\alpha > 0$,*

$$\int\limits_{H} |(F''_{H_\alpha}(x) f_k, f_k)_H|^p \mu(dx) < \infty$$

for some p, $2/3 < p \leq 1$, and $\{f_k\}$ be an orthonormal basis in H, $f_k \in H_\alpha$. If

$$\sum_{k=1}^{n} \int\limits_{H} |(F''_{H_\alpha}(x) f_k, f_k)_H|^p \mu(dx) = O\left(\frac{n^p}{\psi_\varepsilon(n)}\right), \tag{4.1}$$

where $\psi_\varepsilon(n)$ is one of functions

$$(\ln n)^{1+\varepsilon}, \quad \ln n(\ln\ln\ln n)^{1+\varepsilon}, \quad \ldots, \quad \ln n\cdots\underbrace{\ln\cdots\ln}_{m-1} n\,(\underbrace{\ln\cdots\ln}_{m} n)^{1+\varepsilon}$$

for all $\varepsilon > 0$, then

$$\Delta_L F(x) = 0 \quad \text{for almost all } x \in H. \tag{4.2}$$

On the other hand, there exists a function $F(x)$ such that (4.1) is satisfied for $\varepsilon = 0$, but (4.2) is not valid.

Estimate (4.1) (and, therefore, the harmonicity of $F(x)$) does not depend on the choice of the basis from a class of square close bases.

Proof. The Lévy Laplacian (1.3) is the limit of the sequence of arithmetic mean values $\frac{1}{n}\sum_{k=1}^{n}\xi_k(x)$, where $\xi_k(x) = (F''_{H_\alpha}(x)f_k, f_k)_H$.

Any function $\Phi(x)$ on H that is measurable with respect to a σ-algebra $\mathfrak{A}$, is a random variable on the probability space $\{H, \mathfrak{A}, \mu\}$. Its mean value is given by

$$E\,\Phi = \int_H \Phi(x)\,\mu(dx),$$

and the convergence of a sequence of random variables with probability 1 corresponds to the convergence of the sequence of measurable functions almost everywhere on H with respect to μ.

Therefore, $\{\xi_k(x)\}_1^\infty$ is a sequence of arbitrary (dependent) random variables,

$$E|\xi_k|^p < \infty, \quad \text{and} \quad \sum_{k=1}^{n} E\,|\xi_k|^p = O\left(\frac{n^p}{\psi_\varepsilon(n)}\right) \quad \text{for all } \varepsilon > 0.$$

This follows from the theorem of Petrov [108] that $\frac{1}{n}\sum_{k=1}^{n}\xi_k(x) \to 0$ almost surely, i.e.,

$$\Delta_L F(x) = \lim_{n\to\infty}\frac{1}{n}\sum_{k=1}^{n}(F''_{H_\alpha}(x)f_k, f_k)_H = 0$$

for almost all $x \in H$.

Let us show that there exists a function such that the estimates of the theorem are fulfilled for $\varepsilon = 0$, but the function ceases to be harmonic.

Consider

$$F(x) = \sum_{k=1}^{\infty} \int_{-\frac{1}{2}}^{x_k}\int_{-\frac{1}{2}}^{y_k} \varphi_k(\lambda_k z_k)\,dz_k\,dy_k$$

on l_2, where $\varphi_k(\zeta)$ are defined on $\mathbb{R}^1$ and

$$\varphi_k(\zeta) = (2\pi)^{1/2}ke^{\zeta^2/2} \quad \text{for} \quad \zeta \in \left(\frac{1}{2\psi_0(k)}, \frac{k+1}{2k\psi_0(k)}\right),$$

$$\varphi_k(\zeta) = -(2\pi)^{1/2}ke^{\zeta^2/2} \quad \text{for} \quad \zeta \in \left(-\frac{k+1}{2k\psi_0(k)}, -\frac{1}{2\psi_0(k)}\right),$$

and $\varphi_k(\zeta) = 0$ outside these intervals if $k > k_1$; if $k \leq k_1$ then $\varphi_k(\zeta) = 1$ for $\zeta > 0$; and $\varphi_k(\zeta) = -1$ for $\zeta < 0$. Choose the number k_1 to satisfy the condition $(k_1\psi_0(k_1)/(k_1 + 1)) > 1$; $x_k = (x, e_k)_{l_2}$; $\{e_k\}_1^\infty$ is a canonical basis in l_2.

In this case

$$\Delta_L F(x) = \lim_{n\to\infty} \frac{1}{n} \sum_{k=1}^n \varphi_k(\lambda_k x_k).$$

The random variables $\xi_k(x) = \varphi_k(\lambda_k x_k)$ are independent.

For all k, we have $E\,\xi_k = 0$. If $k \leq k_1$, then $E\,|\xi_k| = 1$, and for $k > k_1$ we have $E\,|\xi_k| = 1/\psi_0(k)$. Therefore, if $p = 1$, then

$$\sum_{k=1}^n E\,|\xi_k| = O\left(\frac{n}{\psi_0(n)}\right)$$

and condition (4.1) is satisfied for $\varepsilon = 0$.

For $k > k_1$, the probability

$$P(|\xi_k| \geq k) = \mu\{x \in l_2 : |\xi_k(x)| \geq k\}$$

$$= \frac{2}{\sqrt{2\pi}} \int\limits_{1/2\psi_0(k)}^{k+1/2k\psi_0(k)} e^{-\zeta^2/2}\,d\zeta \geq \frac{1}{\sqrt{2\pi}e^{1/8}k\psi_0(k)},$$

because for $k > k_1$ we have $(k\psi_0(k)/(k + 1)) > 1$, and for

$$\zeta \in \left(\frac{1}{2\psi_0(k)}, \frac{k+1}{2k\psi_0(k)}\right)$$

we have $e^{-\frac{\zeta^2}{2}} \geq e^{-\frac{(k+1)^2}{8k^2\psi_0^2(k)}} \geq e^{-1/8}$. The series $\sum_{k=k_1}^\infty 1/(k\psi_0(k))$ diverges, therefore $\sum_{k=1}^\infty P(|\xi_k| \geq k) = \infty$, and, according to the lemma of Borel–Cantelli, $P(|\xi_k| \geq k \text{ i.m.t.}) = 1$, where i.m.t. means 'infinitely many times'.

Therefore, the relation $\frac{1}{n}\sum_{k=1}^n \xi_k(x) \to 0$ almost surely (i.e. $\Delta_L F(x) = 0$ for almost all $x \in l_2$) is not valid. Indeed, if (4.2) holds, then $\xi_k(x)/k \to 0$ almost surely, which contradicts the fact that $P(|\xi_k| \geq k \text{ i.m.t.}) = 1$.

To prove the final part of the theorem, choose the orthonormal basis $\{g_k\}_1^\infty$ in H, $g_k \in H_\alpha$, to be different from $\{f_k\}_1^\infty$ and square close to $\{f_k\}_1^\infty$, i.e.,

$$\sum_{k=1}^\infty \|g_k - f_k\|_H^2 < \infty.$$

Then for $p \leq 1$

$$\left| \sum_{k=1}^{n} \int_{H} |(F''_{H_\alpha}(x)g_k, g_k)_H|^p \mu(dx) - \sum_{k=1}^{n} \int_{H} |(F''_{H_\alpha}(x)f_k, f_k)_H|^p \mu(dx) \right|$$

$$\leq \sum_{k=1}^{n} \int_{H} |(F''_{H_\alpha}(x)g_k, g_k)_H - (F''_{H_\alpha}(x)f_k, f_k)_H|^p \mu(dx)$$

$$= \sum_{k=1}^{n} \int_{H} |(F''_{H_\alpha}(x)(g_k - f_k), g_k)_H + (F''_{H_\alpha}(x)f_k, g_k - f_k)_H|^p \mu(dx)$$

$$\leq 2c^p \sum_{k=1}^{n} \|g_k - f_k\|_H^p,$$

since $F(x)$ is differentiable with respect to H_α, and hence the operator

$$F''_{H_\alpha}(x) \in \{H_\alpha \to H_{-\alpha}\} : \|F''_{H_\alpha}(x)h\|_{H_{-\alpha}} \leq C(x)\|h\|_{H_\alpha}$$

is bounded, we get

$$|(F''_{H_\alpha}(x)h, g)_H| \leq C(x)\|h\|_{H_\alpha}\|g\|_{H_\alpha},$$

and

$$\int_{H} |(F''_{H_\alpha}(x)h, g)_H| \mu(dx) \leq c\|h\|_H \|g\|_H,$$

$$c = \int_{H} C(x)\mu(dx), \quad h, g \in H_\alpha$$

But

$$\sum_{k=1}^{\infty} \frac{\|g_k - f_k\|_H^p}{[k\psi_\varepsilon(k)]^{\frac{2-p}{2}}} \leq \left(\sum_{k=1}^{\infty} \|g_k - f_k\|_H^2 \right)^{p/2} \left(\sum_{k=1}^{\infty} \frac{1}{k\psi_\varepsilon(k)} \right)^{\frac{2-p}{2}},$$

by the Hölder inequality, and

$$\sum_{k=1}^{\infty} \|g_k - f_k\|_H^2 < \infty, \quad \sum_{k=1}^{\infty} \frac{1}{k\psi_\varepsilon(k)} < \infty \quad (\varepsilon > 0),$$

so, by Kronecker's lemma,

$$\sum_{k=1}^{n} \|g_k - f_k\|_H^p = o\left([n\psi_\varepsilon(n)]^{\frac{2-p}{2}} \right).$$

Hence, taking into account that $[n\psi_\varepsilon(n)]^{\frac{2-p}{2}} = o(n^p/\psi_\varepsilon(n))$ for $p > \frac{2}{3}$ and $\sum_{k=1}^{n} \int_H |(F''_{H_\alpha}(x)f_k, f_k)_H|^p \mu(dx) = O(n^p/\psi_\varepsilon(n))$, we obtain

$\sum_{k=1}^{n} \int_{H} |(F''_{H_\alpha}(x)g_k, g_k)_H|^p \mu(dx) = O(n^p/\psi_\varepsilon(n))$. According to the first statement of this theorem,

$$\Delta_L F(x) = \lim_{n \to \infty} \frac{1}{n} \sum_{k=1}^{n} (F''_{H_\alpha}(x)g_k, g_k)_H = 0$$

for almost all $x \in H$. □

Condition (4.1) of the theorem is in any case not necessary. One can see this by choosing the function

$$F(x) = \frac{1}{6} \sum_{k=1}^{\infty} \lambda_k x_k \lambda_{k+1} x_{k+1}^3$$

on l_2. For this function,

$$\Delta_L F(x) = \lim_{n \to \infty} \frac{1}{n} \sum_{k=1}^{n} \lambda_k x_k \lambda_{k+1} x_{k+1}.$$

The random variables $\xi_k(x) = \lambda_k x_k \lambda_{k+1} x_{k+1}$ are orthogonal, $E\xi_k = 0$, $E\xi_k^2 = 1$, $\sum_{k=1}^{n} E\xi_k^2 = n \leq c(n^2/(\psi_\varepsilon(n)\ln^2 n))$, and it follows from Theorem 4.2 to be presented below that $\Delta_L F(x) = 0$ for almost all $x \in l_2$. At the same time, condition (4.1) is not satisfied, because

$$E|\xi_k| = \int_{l_2} |\lambda_k x_k \lambda_{k+1} x_{k+1}| \mu(dx) = \left(\frac{1}{\sqrt{2\pi}} \int_{-\infty}^{\infty} |y| e^{-y^2/2} \, dy\right)^2 = \frac{2}{\pi},$$

and $\sum_{k=1}^{n} E|\xi_k| = \frac{2}{\pi} n$.

A similar phenomenon can be seen in the case

$$F(x) = \frac{1}{6} \sum_{k=1}^{\infty} \lambda_k x_k^3$$

on l_2. For this function,

$$\Delta_L F(x) = \lim_{n \to \infty} \frac{1}{n} \sum_{k=1}^{n} \lambda_k x_k.$$

The random variables $\xi_k(x) = \lambda_k x_k$ are independent, $E\xi_k = 0$, $E\xi_k^2 = 1$, $\sum_{k=1}^{n} E\xi_k^2 = n \leq c(n^2/\psi_\varepsilon(n))$, and it follows from Theorem 4.3, also presented below, that $\Delta_L F(x) = 0$ for almost all $x \in l_2$. At the same time, condition (4.1) is not satisfied, because

$$E|\xi_k| = \int_{l_2} |\lambda_k x_k| \mu(dx) = \frac{1}{\sqrt{2\pi}} \int_{-\infty}^{\infty} |y| e^{-y^2/2} \, dy = \sqrt{\frac{2}{\pi}},$$

and $\sum_{k=1}^{n} E|\xi_k| = \sqrt{\frac{2}{\pi}} n$.

The reason for this phenomenon is the fact that, for given sums of orthogonal and independent random variables, there exist more precise estimates than for a sum of arbitrary (dependent) random variables. However, in this case the function $F(x)$ has to satisfy additional restrictions.

4.2 Orthogonal and stochastically independent second order derivatives

Theorem 4.2 *Let $F(x)$ be a function twice differentiable with respect to the subspace H_α for some $\alpha > 0$, and $\xi_k(x) = (F''_{H_\alpha} f_k, f_k) \in \mathfrak{L}_2(H, \mu)$, $\{f_k\}_1^\infty$ be an orthonormal basis in H, $f_k \in H_\alpha$. Let, in addition, functions $\xi_k(x)$ satisfy the conditions $(\xi_k, 1)_{\mathfrak{L}_2(H,\mu)} = 0$, $(\xi_j, \xi_k)_{\mathfrak{L}_2(H,\mu)} = 0$ for $j \neq k$, $j, k = 1, 2, \ldots$. If*

$$\sum_{k=1}^n \|\xi_k(x)\|^2_{\mathfrak{L}_2(H,\mu)} = O\left(\frac{n^2}{\psi_\varepsilon(n) \ln^2 n}\right), \tag{4.3}$$

with $\psi_\varepsilon(n)$ determined in Theorem 4.1, then

$$\Delta_L F(x) = 0 \quad \text{for almost all } x \in H. \tag{4.4}$$

On the other hand, there exists a function $F(x)$ such that (4.3) is satisfied for $\varepsilon = 0$ but (4.4) does not hold.

Estimate (4.3) (and, therefore, the harmonicity of $F(x)$) does not depend on the choice of basis from the class of square close bases.

Proof. The Lévy Laplacian (1.3) is the limit of the sequence of arithmetic means $\frac{1}{n} \sum_{k=1}^n \xi_k(x)$. It follows from the conditions of the theorem that $\{\xi_k(x)\}_1^\infty$ is a sequence of orthogonal random variables, with expectation

$$\mathsf{E}\,\xi_k = (\xi_k, 1)_{\mathfrak{L}_2(H,\mu)} = 0, \quad \mathsf{E}\,\xi_k \xi_j = 0 \quad (j \neq k),$$

and variance

$$\text{Var}\,\xi_k = \|\xi_k\|^2_{\mathfrak{L}_2(H,\mu)} < \infty, \quad \text{and} \quad \sum_{k=1}^n \text{Var}\,\xi_k = O\left(\frac{n^2}{\psi_\varepsilon(n) \ln^2 n}\right).$$

According to Petrov's theorem [109], if $\{\zeta_k\}_1^\infty$ is the sequence of orthogonal random variables, $B_n = \sum_{k=1}^n \text{Var}\,\zeta_k \to \infty$, then $\frac{1}{n} \sum_{k=1}^n \zeta_k = o(\sqrt{B_n f(B_n)} \ln n)$ almost surely, where $f(\tau)$ is a positive function on $[\tau_0, \infty)$ such that $\sum_{k=k_0}^\infty (1/kf(k)) < \infty$ (for example, $f(n) = \psi_\varepsilon(n)$, $\varepsilon > 0$). It follows from this theorem that if $\mathsf{E}\,\zeta_k = 0$, $\sum_{k=1}^n \text{Var}\,\zeta_k = O(n^2/\psi_\varepsilon(n) \ln^2 n)$, then $\frac{1}{n} \sum_{k=1}^n \zeta_k \to 0$ almost surely. Therefore, according to the conditions of Theorem 4.2, $\frac{1}{n}\sum_{k=1}^{n}\xi_k \to 0$ almost surely, i.e.,

$$\Delta_L F(x) = \lim_{n\to\infty} \frac{1}{n}\sum_{k=1}^{n}(F''_{H_\alpha}(x)f_k, f_k)_H = 0$$

for almost all $x \in H$.

Now let us show that there exists a function $F(x)$ such that although the conditions of Theorem 4.2 are fulfilled, for $\varepsilon = 0$, $F(x)$ fails to be harmonic.

In the space l_2, consider a function $F(x)$ defined on l_2, by

$$F(x) = \frac{1}{6}\sum_{k=k_0}^{\infty}\sqrt{\frac{k}{\psi_0(k)\ln^2 k}}\prod_{j=1}^{k-1}\lambda_j x_j \lambda_k x_k^3$$

where $x_k = (x_k, e_k)_{l_2}$, $\{e_k\}_1^{\infty}$ is a canonical basis in l_2 and k_0 is chosen so as to ensure that $\underbrace{\ln\cdots\ln}_{m} k_0 > 0$.

Then

$$\Delta_L F(x) = \lim_{n\to\infty} \frac{1}{n}\sum_{k=k_0}^{n}\sqrt{\frac{k}{\psi_0(k)\ln^2 k}}\prod_{j=1}^{k}\lambda_j x_j.$$

The random variables

$$\xi_k(x) = \sqrt{\frac{k}{\psi_0(k)\ln^2 k}}\prod_{j=1}^{k}\lambda_j x_j$$

are orthogonal:

$$\mathrm{E}\,\xi_k = 0, \quad \mathrm{Var}\,\xi_k = \frac{k}{\psi_0(k)\ln^2 k}.$$

Therefore

$$\sum_{k=1}^{n}\mathrm{Var}\,\xi_k = O\left(\frac{n^2}{\psi_0(n)\ln^2 n}\right),$$

and condition (4.3) is satisfied for $\varepsilon = 0$.

At the same time, the relation $\frac{1}{n}\sum_{k=1}^{n}\xi_k(x) \to 0$ for almost all $x \in l_2$ does not hold, because

$$\frac{1}{n}\sum_{k=k_0}^{n}\xi_k(x) = \sum_{k=k_0}^{n}\frac{\xi_k(x)}{k} - \sum_{k=k_0}^{n-1}\frac{1}{k(k+1)}\sum_{j=1}^{k}\xi_j(x)$$

(Abel's transformation), the series $\sum_{k=k_0}^{\infty}\xi_k(x)/k$ diverges, whereas the series $\sum_{k=k_0}^{\infty}1/(k(k+1))\sum_{j=1}^{k}\xi_j(x)$ converges almost everywhere on l_2. Indeed,

$$\sum_{k=k_0}^{\infty}\frac{\xi_k(x)}{k} = \sum_{k=k_0}^{\infty}c_k\hat{\xi}_k(x),$$

where

$$\hat{\xi}_k(x) = \frac{\xi_k(x)}{\|\xi_k(x)\|_{\mathcal{L}_2(H,\mu)}}$$

is a orthonormal system,

$$c_k = \frac{1}{\sqrt{k\psi_0(k)}\,\ln k}$$

is a positive monotone decreasing numerical sequence,

$$\sum_{k=k_0}^{\infty} c_k^2 \ln^2 k = \sum_{k=k_0}^{\infty} \frac{1}{k\psi_0(k)} = \infty,$$

and, by the Men'shov–Rademacher theorem, it follows that this series diverges.

Due to the orthogonality of $\xi_k(x)$ and condition (4.3) at $\varepsilon = 0$, we have

$$\sum_{k=k_0}^{\infty} \frac{1}{k(k+1)} \int_{l_2} \Big|\sum_{j=1}^{k} \xi_j(x)\Big| \mu(dx) \le \sum_{k=k_0}^{\infty} \frac{1}{k(k+1)} \Big[\int_{l_2} \Big(\sum_{j=1}^{k} \xi_j(x)\Big)^2 \mu(dx)\Big]^{1/2}$$

$$= \sum_{k=k_0}^{\infty} \frac{1}{k(k+1)} \Big[\sum_{j=1}^{k} \int_{l_2} \xi_j^2(x)\mu(dx)\Big]^{1/2} \le A \sum_{k=k_0}^{\infty} \frac{1}{(k+1)\sqrt{\psi_0(k)}\,\ln k} < \infty,$$

and the series $\sum_{k=k_0}^{\infty} 1/(k(k+1)) \sum_{j=1}^{k} \xi_j(x)$ converges for almost all $x \in l_2$.

Proof of the last statement of the theorem.

Let $\{g_k\}_1^{\infty}$ be an orthonormal basis in H, $g_k \in H_\alpha$, distinct from $\{f_k\}_1^{\infty}$ and square close to $\{f_k\}_1^{\infty}$. Without loss of generality, assume that

$$\frac{(F''_{H_\alpha}(x)g_n, g_n)_H}{n} \to 0$$

(a necessary condition for $\frac{1}{n}\sum_{k=1}^{n}(F''_{H_\alpha}(x)g_k, g_k)_H \to 0$). Then

$$\Big|\sum_{k=1}^{n} \int_H (F''_{H_\alpha}(x)g_k, g_k)_H^2 \mu(dx) - \sum_{k=1}^{n} \int_H (F''_{H_\alpha}(x)f_k, f_k)_H^2 \mu(dx)\Big|$$

$$= \Big|\sum_{k=1}^{n} \int_H \Big[(F''_{H_\alpha}(x)g_k, g_k)_H - (F''_{H_\alpha}(x)f_k, f_k)_H\Big]$$

$$\times \Big[(F''_{H_\alpha}(x)g_k, g_k)_H + (F''_{H_\alpha}(x)f_k, f_k)_H\Big]\mu(dx)\Big|$$

$$= \Big|\sum_{k=1}^{n} \int_H \Big[(F''_{H_\alpha}(x)(g_k - f_k), g_k)_H + (F''_{H_\alpha}(x)f_k, g_k - f_k)_H\Big]$$

$$\times \Big[(F''_{H_\alpha}(x)g_k, g_k)_H + (F''_{H_\alpha}(x)f_k, f_k)_H\Big]\mu(dx)\Big| \le 4c\sum_{k=1}^{n} k\|g_k - f_k\|_H,$$

because it follows from the condition of differentiability of the function $F(x)$ with respect to subspace H_α that

$$\int_H |(F''_{H_\alpha}(x)h, g)_H|\mu(dx) \le c\|h\|_H\|g\|_H, \quad h, g \in H_\alpha.$$

But

$$\sum_{k=1}^\infty \frac{k\|g_k - f_k\|_H}{\sqrt{k^3\psi_\varepsilon(k)}} \le \left(\sum_{k=1}^\infty \|g_k - f_k\|_H^2\right)^{1/2} \left(\sum_{k=1}^\infty \frac{1}{k\psi_\varepsilon(k)}\right)^{1/2},$$

and

$$\sum_{k=1}^\infty \|g_k - f_k\|_H^2 < \infty, \quad \sum_{k=1}^\infty \frac{1}{k\psi_\varepsilon(k)} < \infty \quad (\varepsilon > 0),$$

so, according to the Kronecker lemma, $\sum_{k=1}^n k\|g_k - f_k\|_H = o(\sqrt{n^3\psi_\varepsilon(n)})$. Hence, taking into account that

$$\sum_{k=1}^n \|(F''_{H_\alpha}(x)f_k, f_k)_H\|_{\mathfrak{L}_2(H,\mu)}^2 = O\left(\frac{n^2}{\psi_\varepsilon(n)\ln^2 n}\right),$$

we have

$$\sum_{k=1}^n \|(F''_{H_\alpha}(x)g_k, g_k)_H\|_{\mathfrak{L}_2(H,\mu)}^2 = O\left(\frac{n^2}{\psi_\varepsilon(n)\ln^2 n}\right).$$

According to the first statement of the theorem,

$$\Delta_L F(x) = \lim_{n\to\infty} \frac{1}{n}\sum_{k=1}^n (F''_{H_\alpha}(x)g_k, g_k)_H = 0$$

for almost all $x \in H$. $\qquad\Box$

Theorem 4.3 *Let function $F(x)$ be twice differentiable with respect to the subspace H_α for some $\alpha > 0$, and $\xi_k(x) = (F''_{H_\alpha}f_k, f_k)_H \in \mathfrak{L}_2(H, \mu)$, where $\{f_k\}_1^\infty$ is an orthonormal basis in H, $f_k \in H_\alpha$. Let also the functions $\xi_k(x)$ satisfy the conditions*

$$\int_H \prod_{k=1}^m g_{p_k}(\xi_{p_k}(x))\mu(dx) = \prod_{k=1}^m \int_H g_{p_k}(\xi_{p_k}(x))\mu(dx) \qquad (4.5)$$

for any finite number of functions $\xi_{p_1}(x), \dots, \xi_{pm}(x)$ and for any bounded and continuous functions $g_{p_1}(\tau), \dots, g_{p_m}(\tau)$, $\tau \in \mathbb{R}^1$. If

$$\sum_{k=1}^n \|\eta_k(x)\|_{\mathfrak{L}_2(H,\mu)}^2 = O\left(\frac{n^2}{\psi_\varepsilon(n)}\right), \qquad (4.6)$$

where $\eta_k(x) = \xi_k(x) - \int_H \xi_k(x)\mu(dx)$, the function $\psi_\varepsilon(n)$ is determined in Theorem 4.1, and

$$\lim_{n\to\infty} \frac{1}{n} \sum_{k=1}^{n} \int_H \xi_k(x)\mu(dx) = 0;$$

then

$$\Delta_L F(x) = 0 \text{ for almost all } x \in H. \tag{4.7}$$

On the other hand, there exists a function $F(x)$ for which at $\varepsilon = 0$ (4.6) is satisfied, but (4.7) does not hold.

Estimate (4.6) (and, therefore, the harmonicity of $F(x)$) does not depend on the choice of the basis from the class of square close bases.

Proof. It follows from the conditions of the theorem that $\{\xi_k(x)\}_1^\infty$ is a sequence of independent random variables, $\{\eta_k(x)\}_1^\infty$ is the sequence given by

$$\eta_k(x) = \xi_k(x) - \mathrm{E}\,\xi_k(x),$$

and the variances satisfy

$$\mathrm{Var}\,\xi_k = \|\eta_k(x)\|^2_{\mathfrak{L}_2(H,\mu)} < \infty, \quad \text{and} \quad \sum_{k=1}^{n} \mathrm{Var}\,\xi_k = O\left(n^2/\psi_\varepsilon(n)\right).$$

According to Theorem 25 from Petrov [110], $\frac{1}{n}\sum_{k=1}^{n}\eta_k \to 0$ almost surely. But, by the above-stated condition of our theorem, $\frac{1}{n}\sum_{k=1}^{n}\mathrm{E}\,\xi_k \to 0$ almost everywhere on $H\frac{1}{n}\sum_{k=1}^{n}\xi_k \to 0$, i.e.,

$$\Delta_L F(x) = \lim_{n\to\infty} \frac{1}{n} \sum_{k=1}^{n} (F''_{H_\alpha}(x)f_k, f_k)_H = 0$$

for almost all $x \in H$. $\qquad\square$

Let us show that there exists a function $F(x)$ such that at $\varepsilon = 0$ (4.6) holds but $F(x)$ is not harmonic.

Consider the function

$$F(x) = \sum_{k=k_1}^{\infty} \int_{-\frac{1}{2k_1\psi_0(k_1)}}^{x_k} \int_{-\frac{1}{2k_1\psi_0(k_1)}}^{y_k} \varphi_k(\lambda_k z_k)\, dz_k dy_k$$

on l_2, where $\varphi_k(\zeta)$ is a function on $\mathbb{R}^1$ such that, for $k > k_1$ we have

$$\varphi_k(\zeta) = (2\pi)^{1/4} k e^{\zeta^2/4} \quad \text{for} \quad \zeta \in \left(0, \frac{1}{2k\psi_0(k)}\right),$$

$$\varphi_k(\zeta) = -(2\pi)^{1/4} k e^{\zeta^2/4} \quad \text{for} \quad \zeta \in \left(-\frac{1}{2k\psi_0(k)}, 0\right),$$

and $\varphi_k(\zeta) = 0$ outside of these intervals; $x_k = (x, e_k)_{l_2}$, $\{e_k\}_1^\infty$ is a canonical basis in l_2; the number k_1 is chosen to be such that $\underbrace{\ln \cdots \ln}_{m} k_1 > 0$.

The Lévy Laplacian of this function has the form

$$\Delta_L F(x) = \lim_{n \to \infty} \frac{1}{n} \sum_{k=k_1}^{n} \varphi_k(\lambda_k x_k).$$

The random variables $\xi_k(x) = \varphi_k(\lambda_k x_k)$ are independent,

$$\mathrm{E}\,\xi_k = 0, \qquad \mathrm{Var}\,\xi_k = \frac{k}{\psi_0(k)}.$$

Therefore

$$\sum_{k=1}^{n} \mathrm{Var}\,\xi_k = O\left(\frac{n^2}{\psi_0(n)}\right),$$

and condition (4.6) is satisfied for $\varepsilon = 0$.

For $k > k_1$, the probability

$$P(|\xi_k| \geq k) = \mu\{x \in l_2 : |\xi_k| \geq k\}$$

$$= \frac{1}{\sqrt{2\pi}} \int_{-\frac{1}{2k\psi_0(k)}}^{\frac{1}{2k\psi_0(k)}} e^{-\zeta^2/2} d\zeta \geq \frac{e^{-1/8k_1^2\psi_0^2(k_1)}}{\sqrt{2\pi}\,k\psi_0(k)},$$

because for $\zeta \in (-1/(2k\psi_0(k)), 1/(2k\psi_0(k)))$ we have $e^{-\zeta^2/2} \geq e^{-1/8k_1^2\psi_0^2(k_1)}$.

The series $\sum_{k=k_1}^\infty 1/(k\psi_0(k))$ diverges, therefore $\sum_{k=k_1}^\infty P(|\xi_k| \geq k) = \infty$, and, according to the Borel–Cantelli lemma $P(|\xi_k| \geq k$ for i.m.t.$) = 1$.

Therefore, the relation $\frac{1}{n}\sum_{k=k_1}^n \xi_k \to 0$ almost surely is not satisfied (i.e., $\Delta_L F(x) = 0$ for almost all $x \in l_2$). Indeed, if (4.7) did hold, then it would be necessary that $\xi_k(x)/k \to 0$ almost surely, which contradicts the equality $P(|\xi_k| \geq k$ for i.m.t.$) = 1$.

The proof of the final part of Theorem 4.3 is similar to the final part of the proof of the Theorem 4.2.

4.3 Translationally non-positive case

One can find functions for which conditions of Theorems 4.1–4.3 are not satisfied but which, nevertheless, are harmonic. To this end one more condition can be useful and will be given below. Notice that this condition does not include the requirement of orthogonality or stochastic independence of second derivatives, and so it is a certain complement to Theorem 4.1.

Let us construct an example of a function for which the conditions of harmonicity of Theorems 4.1–4.3 are not satisfied, but it is harmonic μ-almost everywhere.

Consider the function

$$F(x) = \sum_{k=1}^{\infty} \frac{1}{\ln 2k \lambda_{2k-1} \lambda_{4k-2}} (e^{y_{2k-1} + y_{4k-2}} + e^{-2y_{2k-1}} + e^{-2y_{4k-2}})$$

on l_2, where $y_k = \lambda_k x_k$, $x_k = (x, e_k)_H$, $\{e_k\}_1^{\infty}$ is a canonical basis in l_2.

The Lévy Laplacian of this function is

$$\Delta_L F(x) = \lim_{n \to \infty} \frac{1}{n} \sum_{k=1}^{n} \xi_k(x),$$

where

$$\xi_{p_k} = 0 \quad \text{for} \quad p_k \neq 2k - 1, \quad 4k - 2,$$

$$\xi_{2k-1}(x) = \frac{\lambda_{2k-1}}{\ln 2k \lambda_{4k-2}} (e^{y_{2k-1} + y_{4k-2}} + 4e^{-2y_{2k-1}}), \tag{4.8}$$

$$\xi_{4k-2}(x) = \frac{\lambda_{4k-2}}{\ln 2k \lambda_{2k-1}} (e^{y_{2k-1} + y_{4k-2}} + 4e^{-2y_{4k-2}}), k = 1, 2, \ldots.$$

The condition of harmonicity (4.1) in Theorem 4.1 is not satisfied, because

$$\int_{l_2} |\xi_{2k-1}(x)|^p \mu(dx) = \frac{c\lambda_{2k-1}}{\ln 2k \lambda_{4k-2}},$$

$$\int_{l_2} |\xi_{4k-2}(x)|^p \mu(dx) = \frac{c\lambda_{4k-2}}{\ln 2k \lambda_{2k-1}}.$$

Both the conditions $(\xi_k, \xi_j)_{\mathcal{L}_2(H,\mu)} = 0$ $(j \neq k)$ in Theorem 4.2 and (4.5) of Theorem 4.3 are not satisfied.

Nevertheless, the function $F(x)$ is μ–harmonic almost everywhere. It satisfies the conditions of Theorem 4.4 given below, and hence is harmonic.

Theorem 4.4 *Let function $F(x)$ be twice differentiable with respect to the subspace H_α for some $\alpha > 0$, and $\xi_k(x) = (F''_{H_\alpha} f_k, f_k)_H \in \mathcal{L}_2(H, \mu)$, where $\{f_k\}_1^\infty$ is some orthonormal basis in H, $f_k \in H_\alpha$. If the functions $\xi_k(x)$ satisfy the conditions*

$$\xi_k(x) \geq 0,$$

$$\int_H \xi_j(x)\xi_k(x)\mu(dx) \leq \int_H \xi_j(x)\mu(dx) \int_H \xi_k(x)\mu(dx) \quad \text{for } j \neq k,$$

$$\sup_{k \geq 1} \int_H \xi_k(x)\mu(dx) \leq \infty, \quad \sum_{k=1}^{\infty} \frac{\|\eta_k\|_{\mathcal{L}_2(H,\mu)}^2}{k^2} < \infty,$$

where

$$\eta_k(x) = \xi_k(x) - \int_H \xi_k(x)\mu(dx),$$

and if

$$\lim_{n\to\infty} \frac{1}{n}\sum_{k=1}^{n}\int_H \xi_k(x)\mu(dx) = 0,$$

then

$$\Delta_L F(x) = 0 \quad \text{for almost all } x \in H.$$

Proof. It follows from the conditions of the theorem that $\{\xi_k\}_1^\infty$ is the sequence of arbitrary (dependent) random variables, and

$$\xi_k(x) \geq 0, \quad \mathrm{E}\xi_j\xi_k \leq \mathrm{E}\xi_j\mathrm{E}\xi_k \quad \text{for} \quad j \neq k,$$

$$\sum_{k=1}^{\infty} \frac{\mathrm{Var}\,\xi_k}{k^2} < \infty, \quad \sup_{k\geq 1}\int_H \xi_k(x)\,\mu(dx) < \infty.$$

Note that the Gramm matrix $\|(\hat{\eta}_j, \hat{\eta}_k)_{\mathcal{L}_2(H,\mu)}\|_{j,k=1}^\infty$ of functions $\{\hat{\eta}_k(x)\}_1^\infty$, where $\hat{\eta}_k(x) = \eta_k(x)/\|\eta_k(x)\|_{\mathcal{L}_2(H,\mu)}$, i.e., the matrix of the coefficients of correlation of the random variables $\{\xi_k(x)\}_1^\infty$, is a translationally non-positive matrix[1], since $(\hat{\eta}_j, \hat{\eta}_k)_{\mathcal{L}_2(H,\mu)} = 1$, and by the condition of the theorem

$$\int_H \xi_j(x)\xi_k(x)\,\mu(dx) \leq \int_H \xi_j(x)\,\mu(dx)\int_H \xi_k(x)\,\mu(dx) \quad \text{for} \quad j \neq k$$

so $(\hat{\eta}_j, \hat{\eta}_k)_{\mathcal{L}_2(H,\mu)} \leq 0$ for $j \neq k$, $j, k = 1, 2, \dots$.

According to the Etemadi theorem [35], $\frac{1}{n}\sum_{k=1}^{n}\eta_k(x) \to 0$ almost surely. But, by the assumptions of our theorem, $\frac{1}{n}\sum_{k=1}^{n}\mathrm{E}\xi_k(x) \to 0$, therefore $\frac{1}{n}\sum_{k=1}^{n}\xi_k(x) \to 0$ almost surely, i.e.,

$$\Delta_L F(x) = \lim_{n\to\infty} \frac{1}{n}\sum_{k=1}^{n}(F''_{H_\alpha}(x)f_k, f_k)_H = 0$$

for almost all $x \in H$. □

Let us come back to the above example. The functions $\xi_k(x)$, $k = 1, 2, \dots$, given by (4.8), satisfy the conditions of Theorem 4.4:

$$\xi_{2k-1}(x) > 0, \quad \xi_{4k-2}(x) > 0,$$

$$\xi_{p_k}(x) = 0 \quad \text{for} \quad p_k \neq 2k - 1, 4k - 2,$$

---

[1] The matrix $\|a_{jk}\|_{j,k=1}^\infty$ is called translationally non-positive, if for some $\nu > 0$, $a_{jk} - \nu\delta_{jk} \leq 0$.

so we have $\xi_k(x) \geq 0$;

$$\int_{l_2} \xi_{2k-1}(x)\xi_{4k-2}(x)\,\mu(dx) = \frac{5e^4 + 8e}{\ln^2 2k}$$

$$\leq \int_{l_2} \xi_{2k-1}(x)\,\mu(dx) \int_{l_2} \xi_{4k-2}(x)\,\mu(dx) = \frac{(4e^2 + e)^2}{\ln^2 2k},$$

$$\int_{l_2} \xi_{2i-1}(x)\xi_{4l-2}(x)\,\mu(dx) = \int_{l_2} \xi_{2i-1}(x)\,\mu(dx) \int_{l_2} \xi_{4l-2}(x)\,\mu(dx) \text{ for } i \neq l;$$

therefore

$$\int_{l_2} \xi_j(x)\xi_k(x)\,\mu(dx) \leq \int_{l_2} \xi_j(x)\,\mu(dx) \int_{l_2} \xi_k(x)\,\mu(dx),$$

$$\|\eta_{2k-1}\|^2_{\mathfrak{L}_2(H,\mu)} = \frac{b\lambda^2_{2k-1}}{\ln^2 2k\lambda^2_{4k-2}}, \quad \|\eta_{4k-2}\|^2_{\mathfrak{L}(H,\mu)} = \frac{b\lambda^2_{4k-2}}{\ln^2 2k\lambda^2_{2k-1}},$$

hence

$$\sum_{k=1}^{\infty} \frac{\|\eta_k\|^2_{\mathfrak{L}_2(H,\mu)}}{k^2} < \infty,$$

$$\int_{l_2} \xi_{2k-1}(x)\,\mu(dx) = \frac{c\lambda_{2k-1}}{\ln 2k\lambda_{4k-2}}, \quad \int_{l_2} \xi_{4k-2}(x)\,\mu(dx) = \frac{c\lambda_{4k-2}}{\ln 2k\lambda_{2k-1}}$$

$(b, c$ are constants), so

$$\lim_{n \to \infty} \frac{1}{n} \sum_{k=1}^{n} \int_{l_2} \xi_k(x)\,\mu(dx) = 0.$$

According to Theorem 4.4, $\Delta_L F(x) = 0$ for almost all $x \in l_2$.

5

Linear elliptic and parabolic equations with Lévy Laplacians

The theory of linear equations with Lévy Laplacians is an essential part of infinite-dimensional analysis. Numerous authors have obtained solutions of various kinds of the problems in a variety of functional classes.

5.1 The Dirichlet problem for the Lévy–Laplace and Lévy–Poisson equations

Consider the Dirichlet problem for the Lévy–Laplace equation $\Delta_L U(x) = 0$, and for the Lévy–Poisson equation $\Delta_L U(x) = \Phi(x)$.

Let Ω be a bounded open set in a Hilbert space H (i.e. a domain in H), and $\bar{\Omega} = \Omega \cup \Gamma$ be a domain in H with boundary Γ.

The Dirichlet problem for the equation

$$\Delta_L U(x) = \Phi(x)$$

is to find the function $U(x)$, which satisfies the equation $\Delta_L U(x) = \Phi(x)$ in the domain $\Omega \subset H$ and is equal to some given function $G(x)$ on the boundary Γ of this domain.

We define in H the domain

$$\Omega = \{x \in H : 0 \le Q(x) < 1\}$$

with the boundary

$$\Gamma = \{x \in H : Q(x) = 1\}$$

where $Q(x)$ is a twice differentiable function such that Γ is a convex bounded surface in H, and $\Delta_L Q(x)$ is a constant positive non-zero number. We will call such a domain fundamental.

As simple examples, we mention the ball

$$\bar{\Omega} = \{x \in H : \|x\|_H^2 \le 1\}$$

and the ellipsoid

$$\bar{\Omega} = \{x \in H : (Bx, x)_H \le 1\},$$

where B is a thin operator (i.e., $B = \gamma E + S(x)$, E is the identity operator, and $S(x)$ is a compact operator in H), and $\gamma > 0$.

Consider the Shilov class of functions, i.e. the set of functions of the form

$$\Phi(x) = \phi(x, \|x\|_H^2),$$

where $\phi(x, \xi)$ is a function defined and twice differentiable on $H \times \mathbb{R}^1$, such that $\phi(x, \|x\|_H^2) = \phi(Px, \|x\|_H^2)$, P is a projection on an m-dimensional subspace. In this case,

$$\Delta_L \Phi(x) = 2 \frac{\partial \phi}{\partial \xi} \Big|_{\xi = \|x\|_H^2}.$$

This class contains cylindrical functions $\Psi(x)$, i.e. functions of the form $\Psi(x) = \Psi(Px)$.

Theorem 5.1 (G. E. Shilov) *Let* $G(x) = g(x, \|x\|_H^2)$ *be a function from the Shilov class and* $\bar{\Omega}$ *be a fundamental domain for which the function* $Q(x)$ *has the form*

$$Q(x) = \|x\|_H^2 - \Psi(x) + 1,$$

where $\Psi(x)$ *is a cylindrical positive, twice differentiable function. Then in this class there exists a unique solution of the problem*

$$\Delta_L U(x) = 0 \quad in \quad \Omega, \quad U(x) = G(x) \quad on \quad \Gamma,$$

given by

$$U(x) = g(x, \Psi(x)). \tag{5.1}$$

Proof. The function $g(x, \Psi(x))$ is cylindrical and, therefore, harmonic.

On the boundary Γ, the function $Q(x) = 1$, therefore $U(x) = g(x, \|x\|_H^2) = G(x)$ there.

There exists a unique solution to the problem in the Shilov class. Since the function $U(x)$ is cylindrical for some projector P on $\mathbb{R}^m$ we have

$$U(x) = U(Px) = u((x, a_1)_H, \ldots, (x, a_m)_H), \quad a_1, \ldots, a_m \in H,$$

where $u(\xi_1, \ldots, \xi_m)$ is a function of m variables. If $U(x_0) = c$ at the point $x_0 \in H$ then there exists a hyperplane $\alpha(x_0)$ which goes through the point x_0 such that $U(x)$ remains constant. Note that a hyperplane in H is a manifold of finite co-dimension

$$\alpha(x) = \{x \in H : (b_1, x)_H = c_1, \ldots, (b_m, x) = c_m\}, \quad b_1, \ldots, b_m \in H.$$

On the intersection of $\alpha(x_0)$ and Γ we also have $U(z) = c$, $z \in \Gamma$. Therefore, for any harmonic function $U(x)$ in Ω we have $|U(x)|_{x \in \bar{\Omega}} \leq \sup_{x \in \Gamma} |U(x)|$. It yields the uniqueness of the solution of the Dirichlet problem for the Levy–Laplace equation. $\qquad\square$

Theorem 5.2 (G. E. Shilov) *Let* $\Phi(x) = \phi(x, \|x\|_H^2)$ *be a function from the Shilov class. Under conditions of Theorem 5.1 there exists a unique solution to the problem*

$$\Delta_L U(x) = \Phi(x) \quad in \quad \Omega, \quad U(x) = G(x) \quad on \quad \Gamma,$$

which is equal to

$$U(x) = g(x, \Psi(x)) - \frac{1}{2} \int_{\|x\|_H^2}^{\Psi(x)} \phi(x, \tau) d\tau. \tag{5.2}$$

Proof. From (5.2) by (1.4), we obtain

$$\Delta_L U(x) = \Delta_L g(x, \Psi(x)) - \frac{1}{2} \phi(x, \Psi(x)) \Delta_L \Psi(x)$$

$$+ \frac{1}{2} \phi(x, \|x\|_H^2) \Delta_L \|x\|_H^2 = \phi(x, \|x\|_H^2),$$

since $\Psi(x)$ and $g(x, \Psi(x))$ are cylindrical and, therefore, harmonic, and by (1.2) we have $\Delta_L \|x\|_H^2 = 2$.

On the boundary Γ we have $Q(x) = 1$, therefore we have here $U(x) = g(x, \|x\|_H^2) = G(x)$.

The uniqueness of the solution to the Dirichlet problem for the Levy–Poisson equation follows from the uniqueness of the solution to the Dirichlet problem for the Lévy–Laplace equation. $\qquad\square$

One can extend (5.1) and (5.2) to general fundamental domains

$$\bar{\Omega} = \{x \in H : \quad 0 \leq Q(x) \leq 1, \Delta_L Q(x) = c > 0\}$$

and to the class of functions of the form

$$\Phi(x) = \phi(x, (Ax, x)_H),$$

where $A = \gamma_A E + S_A$, E is the identity operator, and S_A is a compact operator in H, $\phi(x, \xi)$ is a function defined and twice differentiable on $H \times \mathbb{R}^1$, such that $\phi(x, (Ax, x)_H) = \phi(Px, (Ax, x)_H)$, and P is a projection on an m-dimensional subspace. In this case,

$$\Delta_L \Phi(x) = 2\gamma_A \frac{\partial \phi}{\partial \xi}\bigg|_{\xi = (Ax, x)}.$$

We introduce a function

$$T(x) = \frac{1 - Q(x)}{\Delta_L Q(x)},$$

which has the following properties:

$$0 < T(x) \le \frac{1}{\Delta_L Q(x)}, \quad T(x) = 0 \quad \text{at } x \in \Gamma, \quad \Delta_L T(x) = -1 \quad \text{at } x \in \Omega.$$

Theorem 5.3 1. *Let* $G(x) = g(x, (Ax, x)_H)$, $A = \gamma_A E + S_A$, *and the domain* $\bar{\Omega}$ *be fundamental. Then the function*

$$U(x) = g(x, 2\gamma_A T(x) + (Ax, x)_H)) \tag{5.3}$$

is the solution to the problem

$$\Delta_L U(x) = 0 \quad in \quad \Omega, \quad U(x) = G(x) \quad on \quad \Gamma.$$

2. *Let* $\Phi(x) = \phi(x, (Bx, x)_H)$, $B = \gamma_B E + S_B$. *Then the function*

$$U(x) = g(x, 2\gamma_A T(x) + (Ax, x)_H)_H - \frac{1}{2\gamma_B} \int\limits_{(Bx,x)_H}^{2\gamma_B T(x)+(Bx,x)_H} \phi(x, \tau) d\tau \tag{5.4}$$

is a solution to the problem

$$\Delta_L U(x) = \Phi(x) \quad in \quad \Omega, \quad U(x) = G(x) \quad on \quad \Gamma.$$

The proof obviously follows from (1.4) and from the fact that $\Delta_L T(x) = -1$, and, hence

$$\Delta_L [2\gamma_A T(x) + (Ax, x)_H] = 0, \quad \Delta_L [2\gamma_B T(x) + (Bx, x)_H] = 0 \quad in \quad \Omega.$$

In addition, $T\big|_\Gamma = 0$, and, therefore, on Γ we have $U(x) = G(x)$. $\qquad\square$

Consider equations in variational derivatives and functionals $F(x)$ on the Hilbert space of functions $L_2(0, 1)$.

Lemma 5.1 *If a twice strongly differentiable functional* $V(x)$ *attains its minimum (resp. maximum) on* $x_0(t)$, *and* $\Delta_L V(x_0)$ *exists, then*

$$\Delta_L V(x_0) \ge 0 \quad (resp. \ \Delta_L V(x_0) \le 0).$$

The proof obviously follows from the fact that

$$d^2 V(x_0; h) \ge 0 \ (\text{resp. } d^2 V(x_0; h) \le 0) \ \text{ for all } \ h \in L_2(0, 1),$$

at the point of minimum (resp. maximum) x_0, and, therefore by Lemma 1.1,

$$\Delta_L V(x_0) = \mathfrak{M}\{d^2 V(x_0; h)\} \ge 0 \quad (resp. \ \Delta_L V(x_0) \le 0).$$

$\qquad\square$

Lemma 5.2 *Let $W(x)$ be twice strongly differentiable harmonic in Ω, continuous in $\bar{\Omega}$ functional. Then*

$$\sup_{x\in\bar{\Omega}} W(x) = \sup_{x\in\Gamma} W(x), \quad \inf_{x\in\bar{\Omega}} W(x) = \inf_{x\in\Gamma} W(x).$$

Indeed, given $x(t) \in L_2(0, 1)$ define

$$x_\eta(t) = \frac{1}{2\eta} \int\limits_{t-\eta}^{t+\eta} x(s)ds, \quad x(t) = 0 \quad \text{outside of} \quad [0, 1].$$

The function $x_\eta(t)$ is called a Steklov mean of a function $x(t)$.

Let $L_2^\eta(0, 1)$ be the image of $L_2(0, 1)$ under this mapping, and $\bar{\Omega}_\eta = \bar{\Omega} \cap L_2^\eta(0, 1)$, $\Gamma_\eta = \Gamma \cap L_2^\eta(0, 1)$. It is known that $x_\eta(t) \in L_2(0, 1)$, $x_\eta \to x$ in $L_2(0, 1)$ as $\eta \to 0$, and $\bar{\Omega}_\eta$, Γ_η are closed in $L_2(0, 1)$ and compact for each fixed η.

Since $W(x)$ is harmonic in Ω_η and continuous on $\bar{\Omega}_\eta$, it attains its maximum value on Γ_η. Let, on the contrary, $\sup_{x\in\bar{\Omega}} W(x) = W(x_0) = M_\eta > m_\eta = \sup_{x\in\Gamma_\eta} W(x)$. We set

$$V(x) = W(x) + \frac{M_\eta - m_\eta}{2d} \int\limits_0^1 x^2(s)ds,$$

where

$$d = \sup_{x\in\Gamma_\eta} \int\limits_0^1 x^2(s)ds,$$

and show that it attains its maximum at a certain point $x_0 \in \Omega_\eta$. Let $\sup_{x\in\Gamma_\eta} V(x) = V(x_\Gamma)$, $x_\Gamma \in \Gamma_\eta$. Then we have

$$V(x_\Gamma) = W(x_\Gamma) + \frac{M_\eta - m_\eta}{2d} \int\limits_0^1 x^2(s)ds \le m_\eta + \frac{M_\eta - m_\eta}{2}$$

$$= M_\eta - \frac{M_\eta - m_\eta}{2} < M_\eta.$$

But for all $x \in \bar{\Omega}_\eta$, $V(x) \ge W(x)$, therefore $\sup_{x\in\bar{\Omega}_\eta} V(x) \ge \sup_{x\in\bar{\Omega}_\eta} W(x) = M_\eta$. This means that $x_0 \in \Omega_\eta$. By Lemma 5.1, $\Delta_L V(x_0) \le 0$. Substituting $V(x)$ here and taking into account that $\Delta_L W(x_0) = 0$ (because $x_0 \in \Omega_\eta$), we obtain that $M_\eta \le m_\eta$, which is a contradiction. Thus, $\sup_{x\in\bar{\Omega}_\eta} W(x) = \sup_{x\in\Gamma_\eta} W(x)$.

Since

$$\overline{\lim}_{\eta\to 0} \sup_{x\in\bar{\Omega}_\eta} W(x) = \sup_{x\in\bar{\Omega}} W(x), \quad \overline{\lim}_{\eta\to 0} \sup_{x\in\Gamma_\eta} W(x) = \sup_{x\in\Gamma} W(x),$$

we have

$$\sup_{x\in\bar{\Omega}} W(x) = \sup_{x\in\Gamma} W(x).$$

Similarly we obtain $\inf_{x\in\bar{\Omega}} W(x) = \inf_{x\in\Gamma} W(x)$. $\qquad\square$

Theorem 5.4 *The solution to the Dirichlet problem for the equation $\Delta_L U(x) = 0$ is unique in the class of functionals $U(x)$ that are twice strongly differentiable in Ω, continuous in $\bar{\Omega}$ and such that $\Delta_L U(x)$ exists.*

Proof. Suppose that under the conditions of the theorem

$$\Delta_L U_1(x) = 0, \ \Delta_L U_2(x) = 0 \ \text{ in } \ \Omega, \ \text{ and } \ U_1\big|_\Gamma = U_2\big|_\Gamma = G(x).$$

Then for the functional $W(x) = U_1(x) - U_2(x)$ we have

$$\Delta_L W(x) = 0 \ \text{ in } \ \Omega, \ \ W\big|_\Gamma = 0.$$

By Lemma 5.2, $W(x) = 0$, and so $U_1(x) \equiv U_2(x)$. $\qquad\square$

Theorem 5.5 *The solution to the Dirichlet problem for the equation $\Delta_L U(x) = \Phi(x)$ is unique in the class of functionals $U(x)$ that are twice strongly differentiable in Ω, continuous in $\bar{\Omega}$ and such that $\Delta_L U(x)$ exists.*

Proof. Suppose that under the conditions of the theorem

$$\Delta_L U_1(x) = \Phi(x), \quad \Delta_L U_2(x) = \Phi(x) \ \text{ in } \ \Omega, \ \text{ and } \ U_1\big|_\Gamma = U_2\big|_\Gamma = 0.$$

Then for the functional $W(x) = U_1(x) - U_2(x)$ we have

$$\Delta_L W(x) = 0 \ \text{ in } \Omega, \ \ W\big|_\Gamma = 0.$$

By Lemma 5.2, $W(x) = 0$, and so $U_1(x) \equiv U_2(x)$. $\qquad\square$

Consider the class of Gâteaux functionals; that is, the set of functionals of the form

$$F(x) = \int_0^1 \cdots \int_0^1 f(x(t_1), \ldots, x(t_N); t_1, \ldots, t_N) dt_1 \cdots dt_N,$$

where $f(\xi_1, \ldots, \xi_N; t_1, \ldots, t_N)$ is a continuous function of $2N$ variables

$(-\infty < \xi_k < \infty, 0 \le t_k \le 1, k = 1, 2, \ldots, N)$, and

$$f(\xi_1, \ldots, \xi_N; t_1, \ldots, t_N) = O\left(b \sum_{k=1}^{N} \xi_k^2\right), \quad 0 < b < \frac{\Delta_L Q}{4}.$$

Theorem 5.6 (E. M. Polishchuk) *Assume that*

$$G(x) = \int_0^1 \cdots \int_0^1 g(x(t_1), \ldots, x(t_N); t_1, \ldots, t_N) dt_1 \cdots dt_N,$$

belongs to the Gâteaux class, the domain $\bar{\Omega}$ is fundamental and $dQ(x; h)$. $d^2 Q(x, h)$ *have normal form. Then the functional*

$$U(x) = \frac{1}{(4\pi T(x))^{N/2}} \int_0^1 \cdots \int_0^1 dt_1 \cdots dt_N$$

$$\times \int_{-\infty}^{\infty} \cdots \int_{-\infty}^{\infty} g(\zeta_1, \ldots, \zeta_N; t_1, \ldots, t_N) e^{-\sum_{k=1}^{N} \frac{(x(t_k) - \zeta_k)^2}{4T(x)}} d\zeta_1 \cdots d\zeta_N, \quad (5.5)$$

where $T(x) = (1 - Q(x))/(\Delta_L Q(x))$, solves the problem

$$\Delta_L U(x) = 0 \quad in \quad \Omega, \quad U(x) = G(x) \quad on \quad \Gamma.$$

Proof. Rewrite (5.5) in the form

$$U(x) = \int_0^1 \cdots \int_0^1 u(x(t_1), \ldots, u(t_N); t_1, \ldots, t_N, T(x)) dt_1 \cdots dt_N, \quad (5.6)$$

where

$$u(\xi_1, \ldots, \xi_N; t_1, \ldots, t_N; \tau) = \frac{1}{(4\pi\tau)^{N/2}} \int_{-\infty}^{\infty} \cdots \int_{-\infty}^{\infty} g(\zeta_1, \ldots, \zeta_N; t_1, \ldots, t_N)$$

$$\times \exp\left\{-\sum_{k=1}^{N} \frac{(\xi_k - \zeta_k)^2}{4\tau}\right\} d\zeta_1 \cdots d\zeta_N$$

is a solution to the Cauchy problem

$$\frac{\partial u}{\partial \tau} = \sum_{k=1}^{N} \frac{\partial^2 u}{\partial \xi_k^2} \quad (0 < \tau \le \frac{1}{\Delta_L Q}, -\infty < \xi_k < \infty),$$

$$u\Big|_{\tau=0} = g(\xi_1, \ldots, \xi_N; t_1, \ldots, t_N). \quad (5.7)$$

Let us derive the expression for the second differential of (5.6). Since we can differentiate under the integral sign we obtain

$$d^2 U(x; h) = \int_0^1 \cdots \int_0^1 \left\{ \sum_{j,k=1}^N \frac{\partial^2 u}{\partial \xi_j \partial \xi_k} h(t_j) h(t_k) dt_j dt_k \right.$$

$$+ 2 \sum_{k=1}^N \frac{\partial^2 u}{\partial \xi_k \partial \tau} h(t_k) dT(x; h)$$

$$\left. + \frac{\partial^2 u}{\partial \tau^2} \left[dT(x; h) \right]^2 + \frac{\partial u}{\partial \tau} d^2 T(x; h) \right\} dt_1 \cdots dt_N.$$

By virtue of Lemma 1.1, taking into account that the differentials of $T(x)$ have normal form (because according to the condition of the theorem, the differentials of $Q(x)$ have normal form) we obtain

$$\Delta_L U(x) = \int_0^1 \cdots \int_0^1 \left\{ \sum_{k=1}^N \frac{\partial^2 u}{\partial \xi_k^2} + \frac{\partial u}{\partial \tau} \Delta_L T(x) \right\} dt_1 \cdots dt_N.$$

But $\Delta_L T(x) = -1$, therefore

$$\Delta_L U(x) = \int_0^1 \cdots \int_0^1 \left\{ \sum_{k=1}^N \frac{\partial^2 u}{\partial \xi_k^2} - \frac{\partial u}{\partial \tau} \right\} dt_1 \cdots dt_N.$$

Since $u(\xi_1, \ldots, \xi_N; t_1, \ldots, t_N; \tau)$ is a solution to the Cauchy problem (5.7) then $\Delta_L U(x) = 0$ in the domain Ω.

On the boundary Γ, we have $T(x) = 0$. Therefore, on this boundary

$$U(x) = \int_0^1 \cdots \int_0^1 u(x(t_1), \ldots, x(t_N); t_1, \ldots, t_N, 0) dt_1 \cdots dt_N$$

$$= \int_0^1 \cdots \int_0^1 g(x(t_1), \ldots, x(t_N); t_1, \ldots, t_N) dt_1 \cdots dt_N = G(x).$$

$\square$

Theorem 5.7 (E. M. Polishchuk) *Let*

$$\Phi(x) = \int_0^1 \cdots \int_0^1 \varphi(x(t_1), \ldots, x(t_N); t_1, \ldots, t_N) dt_1 \cdots dt_N$$

be a functional from the Gâteaux class. Under the conditions of Theorem 5.6, the functional

$$U(x) = \frac{1}{(4\pi T(x))^{N/2}} \int_0^1 \cdots \int_0^1 dt_1 \cdots dt_N$$

$$\times \int_{-\infty}^{\infty} \cdots \int_{-\infty}^{\infty} g(\zeta_1, \ldots, \zeta_N; t_1, \ldots, t_N) e^{-\sum_{k=1}^{N} \frac{(x(t_k)-\zeta_k)^2}{4T(x)}} d\zeta_1 \cdots d\zeta_N$$

$$- \int_0^{T(x)} \left[\int_0^1 \cdots \int_0^1 dt_1 \ldots dt_N \int_{-\infty}^{\infty} \cdots \int_{-\infty}^{\infty} \frac{\varphi(\zeta_1, \ldots, \zeta_N; t_1, \ldots, t_N)}{[4\pi(T(x)-t)]^{N/2}} \right.$$

$$\left. \times e^{-\sum_{k=1}^{N} \frac{(x(t_k)-\zeta_k)^2}{4(T(x)-t)}} d\zeta_1 \cdots d\zeta_N \right] dt \qquad (5.8)$$

is a solution to the problem

$$\Delta_L U(x) = \Phi(x) \quad in \quad \Omega, \quad U(x) = G(x) \quad on \quad \Gamma.$$

Proof. Rewrite (5.8) in the form

$$U(x) = \int_0^1 \cdots \int_0^1 \left[u(x(t_1), \ldots, x(t_N); t_1, \ldots, t_N; T(x)) \right.$$

$$\left. + v(x(t_1), \ldots, x(t_N); t_1, \ldots, t_N; T(x)) \right] dt_1 \cdots dt_N,$$

where $u(\xi_1, \ldots, \xi_N; t_1, \ldots, t_N; \tau)$ is a solution to the Cauchy problem (5.7), and

$$v(\xi_1, \ldots, \xi_N; t_1, \ldots, t_N; \tau)$$

$$= - \int_0^{\tau} \left[\int_{-\infty}^{\infty} \cdots \int_{-\infty}^{\infty} \frac{\varphi(\zeta_1, \ldots, \zeta_N; t_1, \ldots, t_N)}{[4\pi(\tau-t)]^{N/2}} \right.$$

$$\left. \times e^{-\sum_{k=1}^{N} \frac{(\xi_k-\zeta_k)^2}{4(\tau-t)}} d\zeta_1 \cdots d\zeta_N \right] dt$$

is a solution to the Cauchy problem

$$\sum_{k=1}^{N} \frac{\partial^2 v}{\partial \xi_k^2} - \frac{\partial v}{\partial \tau} = \varphi(\xi_1, \ldots, \xi_N; t_1, \ldots, t_N)$$

$$\left(0 < \tau \le \frac{1}{\Delta_L Q}, \quad -\infty < \xi_k < \infty\right),$$

$$v\Big|_{\tau=0} = 0.$$

The final part of the proof is similar to the proof of Theorem 5.6. $\qquad \square$

Note that solutions (5.5) and (5.8) belong to the uniqueness class mentioned in Theorems 5.4 and 5.5.

Let us continue our study of the class of Gâteaux functionals. The Dirichlet problem for the Lévy–Laplace and Lévy–Poisson equations in the class of Gâteaux functionals is connected with the finite-dimensional heat equation (Theorems 5.6 and 5.7). At the same time, the fundamental solution of the heat equation gives rise to the Wiener measure. This lets us obtain the solution of the Dirichlet problem using the integral with respect to the Wiener measure.

Let $\hat{L}_2(0, 1)$ be the space of functions $x(t)$, square integrable on $[0, 1]$ and satisfying the condition $(x, 1)_{L_2(0,1)} = 0$; $\hat{L}_2(0, 1)$ is a subspace of $L_2(0, 1)$. In $\hat{L}_2(0, 1)$, one can define the Wiener measure – the Gaussian measure with the zero mean value and the correlational operator

$$Kx(s) = \frac{1}{2}\int_0^1 \min(t, s)x(s)ds - \frac{1}{2}\int_0^1\int_0^1 \min(\xi, s)x(s)dsd\xi$$

(see Section 1.3).

Theorem 5.8 *Let*

$$G(x) = \int_0^1 \cdots \int_0^1 g(x(t_1), \ldots, x(t_N); t_1, \ldots, t_N)dt_1 \cdots dt_N$$

be a functional from the Gâteaux class. Let $g(\xi_1, \ldots, \xi_N; t_1, \ldots, t_N)$ be twice continuously differentiable in $\xi_1, \ldots, \xi_N$ and, together with its derivatives, integrable in $\mathbb{R}^N$ with respect to the measure having density $(1/(2\pi)^{N/2})e^{-\frac{1}{2}\sum_{i=1}^N \xi_i^2}$. Let in addition the domain $\hat{\Omega}$ be fundamental, and $dQ(x; h)$ and $d^2Q(x; h)$ have normal form. Then there exists a unique solution to the problem

$$\Delta_L U(x) = 0 \quad in \quad \Omega, \quad U(x) = G(x) \quad on \quad \Gamma,$$

which is equal to

$$U(x) = \int_{\hat{L}_2(0,1)}\int_0^1 \cdots \int_0^1 g(x(t_1) + 2\sqrt{T(x)}\, Y(t_1), \ldots, x(t_N)$$
$$+ 2\sqrt{T(x)}\, Y(t_N); t_1, \ldots, t_N)dt_1 \cdots dt_N \mu_w(dy), \quad (5.9)$$

where

$$T(x) = \frac{1 - Q(x)}{\Delta_L Q(x)}, \quad Y(t_k) = \frac{y(\sigma_{t_k}) - y(\sigma_{t_{k-1}})}{\sqrt{t_k}}; \quad \sigma_{t_k} = \sum_{i=1}^k t_i, \quad \sigma_{t_0} = 0.$$

Proof. Let $\hat{C}_0(0, 1)$ denote the set of all continuous functions from $\hat{L}_2(0, 1)$. There exists a one-to-one correspondence between $C_0(0, 1)$ and $\hat{C}_0(0, 1)$ given by

$$z(t) = y(t) - y(0), \quad y(t) = z(t) - \int_0^1 z(s)ds$$

$$(y(t) \in C_0(0, 1), \quad z(t) \in \hat{C}_0(0, 1)).$$

If a functional $F(y)$ is integrable in the Wiener measure on $\hat{L}_2(0, 1)$ and $\Phi(z) = F(z - \int_0^1 z(s)ds)$ is integrable with respect to the classical Wiener measure on $C_0(0, 1)$, then

$$\int_{\hat{C}_0} F(y)\mu_w(dy) = \int_{C_0} \Phi(z)\mu_w(dz).$$

The space $\hat{C}_0(0, 1)$ has full measure in $\hat{L}_2(0, 1)$ hence

$$\int_{\hat{L}_2(0,1)} F(y)\mu_w(dy) = \int_{C_0(0,1)} \Phi(z)\mu_w(dz).$$

Since

$$y(\sigma_{t_k}) - y(\sigma_{t_{k-1}}) = \left[z(\sigma_{t_k}) - \int_0^1 z(s)ds \right] - [z(\sigma_{t_{k-1}}) - \int_0^1 z(s)ds]$$

$$= z(\sigma_{t_k}) - z(\sigma_{t_{k-1}}),$$

we can rewrite (5.9) in the form

$$U(x) = \int_{C_0(0,1)} \int_0^1 \cdots \int_0^1 g(x(t_1) + 2\sqrt{T(x)}\, Z(t_1), \ldots, x(t_N)$$

$$+ 2\sqrt{T(x)}\, Z(t_N); t_1, \ldots, t_N)dt_1 \cdots dt_N \mu_w(dz),$$

where $Z(t_k) = (z(\sigma_{t_k}) - z(\sigma_{t_{k-1}})/\sqrt{t_k})(z(t) \in C_0(0, 1))$.

Let us find the second differential of the functional $U(x)$. By differentiating under the integral sign we obtain

$$d^2U(x; h)) = \int_{C_0(0,1)} \int_0^1 \cdots \int_0^1 \left\{ \sum_{j,k=1}^N \frac{\partial^2 g(\Xi_x(t_1), \ldots, \Xi_x(t_N); t_1, \ldots, t_N)}{\partial \xi_j \partial \xi_k} \right.$$

$$\times \left[h(t_j) + Z(t_j)\frac{\delta T(x; h)}{\sqrt{T(x)}} \right]\left[h(t_k) + Z(t_k)\frac{\delta T(x; h)}{\sqrt{T(x)}} \right]$$

$$+ \sum_{k=1}^{N} \frac{\partial g(\Xi_x(t_1), \ldots, \Xi_x(t_N); t_1, \ldots, t_N)}{\partial \xi_k} Z(t_k)$$

$$\times \left[\frac{\delta^2 T(x; h)}{\sqrt{T(x)}} - \frac{[\delta T(x; h)]^2}{2 T^{3/2}(x)} \right] \Big\} dt_1 \cdots dt_N \mu_w(dz),$$

where $\Xi_x(t_k) = x(t_k) + 2\sqrt{T} \, Z(t_k)$.

By virtue of Lemma 1.1 taking into account that the differentials of $T(x)$ have normal form (because by the condition of the theorem the differentials of $Q(x)$ have normal form) we obtain

$$\Delta_L U(x) = \int\limits_{C_0(0,1)} \int\limits_0^1 \cdots \int\limits_0^1 \sum_{k=1}^{N} \left[\frac{\partial^2 g(\Xi_x(t_1), \ldots, \Xi_x(t_N); t_1, \ldots, t_N)}{\partial \xi_k^2} \right.$$

$$\left. + \frac{\Delta_L T(x)}{\sqrt{T(x)}} Z(t_k) \frac{\partial g(\Xi_x(t_1), \ldots, \Xi_x(t_N); t_1, \ldots, t_N)}{\partial \xi_k} \right] dt_1 \cdots dt_N \mu_w(dz).$$

But $\Delta_L T(x) = -1$, therefore

$$\Delta_L U(x) = \int\limits_{C_0(0,1)} \int\limits_0^1 \cdots \int\limits_0^1 \sum_{k=1}^{N} \left[\frac{\partial^2 g(\Xi_x(t_1), \ldots, \Xi_x(t_N); t_1, \ldots, t_N)}{\partial \xi_k^2} \right.$$

$$\left. - \frac{1}{\sqrt{T(x)}} Z(t_k) \frac{\partial g(\Xi_x(t_1), \ldots, \Xi_x(t_N); t_1, \ldots, t_N)}{\partial \xi_k} \right] dt_1 \cdots dt_N \mu_w(dz).$$

By changing the order of integration, we obtain

$$\Delta_L U(x) = \sum_{j,k=1}^{N} \int\limits_0^1 \cdots \int\limits_0^1 \int\limits_{C_0(0,1)} \left[\frac{\partial^2 g(\Xi_x(t_1), \ldots, \Xi_x(t_N); t_1, \ldots, t_N)}{\partial \xi_k^2} \right.$$

$$\left. - \frac{Z(t_k)}{\sqrt{T(x)}} \frac{\partial g(\Xi_x(t_1), \ldots, \Xi_x(t_N); t_1, \ldots, t_N)}{\partial \xi_k} \right] dt_1 \cdots dt_N \mu_w(dz)$$

$$= \sum_{j,k=1}^{N} \int\limits_0^1 \cdots \int\limits_0^1 I_k(x(t_1), \ldots, x(t_N); t_1, \ldots, t_N) dt_1 \cdots dt_N,$$

where

$$I_k(x(t_1), \ldots x(t_N); t_1, \ldots, t_N) = \int\limits_{C_0(0,1)} \left[\frac{\partial^2 g(\Xi_x(t_1), \ldots, \Xi_x(t_N); t_1, \ldots, t_N)}{\partial \xi_k^2} \right.$$

$$\left. - \frac{Z(t_k)}{\sqrt{T(x)}} \frac{\partial g(\Xi_x(t_1), \ldots, \Xi_x(t_N); t_1, \ldots, t_N)}{\partial \xi_k} \right] dt_1 \cdots dt_N \mu_w(dz).$$

Let us compute the Wiener integral of the functional concentrated at points $\sigma_{t_0}, \sigma_{t_1}, \ldots \sigma_{t_N}$.

If the functional $F(x) = f(z(\tau_1), \ldots, z(\tau_m))$, where $f(\xi_1, \ldots, \xi_m)$ is a function on $\mathbb{R}^m$, and $0 < \tau_1 < \tau_2 < \cdots < \tau_m \leq 1$, then

$$\int_{C_0(0,1)} f(z(\tau_1), \ldots, z(\tau_m))\mu_w(dz)$$

$$= \frac{\pi^{-m/2}}{[\prod_{i=1}^{m}(\tau_i - \tau_{i-1})]^{1/2}} \int_{R^m} f(z_1, \ldots, z_m)e^{-\sum_{i=1}^{m}\frac{(z_i-z_{i-1})^2}{\tau_i-\tau_{i-1}}} dz_1 \cdots dz_m.$$

Since $\sigma_{t_0} < \sigma_{t_1} < \sigma_{t_2} < \cdots < \sigma_{t_N}$ (here $\sigma_{t_0} = 0, \sigma_{t_1} = t_1, \sigma_{t_2} = t_1 + t_2, \ldots, \sigma_{t_N} = \sum_{i=1}^{N} t_i$), we have

$$I_k(x(t_1), \ldots x(t_N); t_1, \ldots, t_N)$$

$$= \frac{1}{(\pi^N \prod_{i=1}^{N} t_i)^{1/2}} \int_{-\infty}^{\infty} \cdots \int_{-\infty}^{\infty} \left[\frac{\partial^2 g(\hat{\Xi}_x(t_1), \ldots, \hat{\Xi}_x(t_N); t_1, \ldots, t_N)}{\partial \xi_k^2} \right.$$

$$\left. - \frac{1}{\sqrt{T(x)}} \frac{z_k - z_{k-1}}{\sqrt{t_k}} \frac{\partial g(\hat{\Xi}_x(t_1), \ldots, \hat{\Xi}_x(t_N); t_1, \ldots, t_N)}{\partial \xi_k} \right]$$

$$\times e^{-\sum_{i=1}^{N}\frac{(z_i-z_{i-1})^2}{t_i}} dz_1 \cdots dz_N,$$

where $\hat{\Xi}_x(t_k) = x(t_k) + 2\sqrt{T}((z_k - z_{k-1})/\sqrt{t_k})$.

Let us change z_k to ζ_k via $(z_k - z_{k-1})/\sqrt{t_k} = \zeta_k/\sqrt{2}$; we obtain

$$I_k(x(t_1), \ldots x(t_N); t_1, \ldots, t_N)$$

$$= \frac{1}{(2\pi)^{N/2}} \int_{-\infty}^{\infty} \cdots \int_{-\infty}^{\infty} \left[\frac{\partial^2 g(\breve{\Xi}_x(t_1), \ldots, \breve{\Xi}_x(t_N); t_1, \ldots, t_N)}{\partial \xi_k^2} \right.$$

$$\left. - \frac{\zeta_k}{\sqrt{2T(x)}} \frac{\partial g(\breve{\Xi}_x(t_1), \ldots, \breve{\Xi}_x(t_N); t_1, \ldots, t_N)}{\partial \xi_k} \right] e^{-\frac{1}{2}\sum_{i=1}^{N}\zeta_i^2} d\zeta_1 \cdots d\zeta_N$$

$$= \frac{1}{(2\pi)^{N/2}} \int_{-\infty}^{\infty} \cdots \int_{-\infty}^{\infty} \frac{\partial^2 g(\breve{\Xi}_x(t_1), \ldots, \breve{\Xi}_x(t_N); t_1, \ldots, t_N)}{\partial \xi_k^2} e^{-\frac{1}{2}\sum_{i=1}^{N}\zeta_i^2} d\zeta_1 \cdots d\zeta_N$$

$$+ \frac{1}{(2\pi)^{N/2}} \int_{-\infty}^{\infty} \cdots \int_{-\infty}^{\infty} \frac{1}{\sqrt{2T(x)}} \frac{\partial g(\breve{\Xi}_x(t_1), \ldots, \breve{\Xi}_x(t_N); t_1, \ldots, t_N)}{\partial \xi_k}$$

$$\times \frac{\partial}{\partial \zeta_k} e^{-\frac{1}{2}\sum_{i=1}^{N}\zeta_i^2} d\zeta_1 \cdots d\zeta_N,$$

where $\breve{\Xi}_x(t_k) = x(t_k) + \sqrt{2T}\zeta_k$.

Integrating by parts (in the second summand) we obtain that $I_k = 0$. Therefore, $\Delta_L U(x) = 0$ in the domain Ω.

On the boundary Γ, the functional $U(x)$ is equal to $G(x)$, because on Γ, we have $T(x) = 0$, and the Wiener measure of the whole space $\hat{L}_2(0,1)$ is equal to 1.

In addition, $U(x)$ lies in the uniqueness class indicated in Theorem 5.4. $\square$

Theorem 5.9 *Let*

$$\Phi(x) = \int_0^1 \cdots \int_0^1 \varphi(x(t_1), \ldots, x(t_N); t_1, \ldots, t_N) dt_1 \cdots dt_N$$

be a functional from the Gâteaux class. Let $\varphi(\xi(t_1), \ldots, \xi(t_N); t_1, \ldots, t_N)$ *be twice continuously differentiable in* $\xi_1, \ldots, \xi_N$ *and, together with its derivatives, integrable in* $\mathbb{R}^N$ *with respect to the measure having the density* $(1/(2\pi)^{N/2})e^{-\frac{1}{2}\sum_{i=1}^N \xi_i^2}$. *Then under the conditions of Theorem 5.8. there exists the unique solution to the problem*

$$\Delta_L U(x) = \Phi(x) \quad in \quad \Omega, \quad U(x) = G(x) \quad on \quad \Gamma,$$

which is equal to

$$U(x) = \int_{\hat{L}_2(0,1)} \int_0^1 \cdots \int_0^1 g(x(t_1) + 2\sqrt{T(x)}Y(t_1), \ldots, x(t_N)$$

$$+ 2\sqrt{T(x)}Y(t_N); t_1, \ldots, t_N) dt_1 \cdots dt_N \mu_w(dy)$$

$$- \int_0^{T(x)} \int_{\hat{L}_2(0,1)} \int_0^1 \cdots \int_0^1 \varphi(x(t_1) + 2\sqrt{T(x) - \xi}Y(t_1), \ldots, x(t_N)$$

$$+ 2\sqrt{T(x) - \xi}\, Y(t_N); t_1, \ldots, t_N) dt_1 \ldots dt_N \mu_w(dy) d\xi.$$

Proof. If one has the solution to the Dirichlet problem for the Lévy–Laplace equation (Theorem 5.8), it is sufficient to show that the solution of the Dirichlet problem with homogeneous boundary conditions for the Lévy–Poisson equation $\Delta_L V(x) = \Phi(x)$ in Ω, $V(x) = 0$ on Γ, is equal to

$$V(x) = -\int_0^{T(x)} \int_{\hat{L}_2(0,1)} \int_0^1 \cdots \int_0^1 \varphi(x(t_1) + 2\sqrt{T(x) - \xi}Y(t_1), \ldots, x(t_N)$$

$$+ 2\sqrt{T(x) - \xi}Y(t_N); t_1, \ldots, t_N) dt_1 \cdots dt_N \mu_w(dy) d\xi.$$

Computing the second differential and the Lévy Laplacian of the functional $V(x)$ (the scheme of computation does not differ from that given in the proof of Theorem 5.8), we obtain

$$
\Delta_L V(x) = -\int_0^1 \cdots \int_0^1 \varphi(x(t_1), \ldots, x(t_N); t_1, \ldots, t_N) dt_1 \cdots dt_N \Delta_L T(x)
$$

$$
-\int_0^{T(x)} \int_{\hat{L}_2(0,1)} \int_0^1 \cdots \int_0^1 \sum_{k=1}^N \Big[\frac{\partial^2 \varphi(\Xi_{x,\xi}(t_1), \ldots, \Xi_{x,\xi}(t_N); t_1, \ldots, t_N)}{\partial \xi_k^2}
$$

$$
+ \frac{\Delta_L T(x)}{\sqrt{T(x) - \xi}} Y(t_k)
$$

$$
\times \frac{\partial \varphi(\Xi_{x,\xi}(t_1), \ldots, \Xi_{x,\xi}(t_N); t_1, \ldots, t_N)}{\partial \xi_k} \Big] dt_1 \cdots dt_N \mu_w(dy) d\xi,
$$

where $\Xi_{x,\xi}(t_k) = x(t_k) + 2\sqrt{T - \xi} Y(t_k)$.

But $\Delta_L T(x) = -1$, and the second summand is equal to zero (see the proof of Theorem 5.8). Therefore $\Delta_L V(x) = \Phi(x)$ in the domain Ω.

On the boundary Γ the functional $V(x) = 0$, because on Γ we have $T(x) = 0$, and the Wiener measure of the whole space $\hat{L}_2(0, 1)$ is equal to 1.

In addition, $V(x)$ lies in the uniqueness class indicated in Theorem 5.5. $\square$

Consider the class of functionals which can be represented in the form of a series in terms of orthogonal polynomials $\mathcal{P}_0, \mathcal{P}_{mq}(x)$, with $m, q = 1, 2, \ldots$, introduced in Section 2.2.

Theorem 5.10 *Let $G(x)$, be a functional defined and continuous on $L_2(0, 1)$, and which admits the representation in the form of a series in the polynomials $\mathcal{P}_0, \mathcal{P}_{mq}(x)$, $m, q = 1, 2, \ldots$ absolutely and uniformly converging on $\bar{\Omega}$. Let the domain $\bar{\Omega}$ be fundamental, and $dQ(x; h)$ and $d^2 Q(x; h)$ have normal form. Then there exists a unique solution of the problem*

$$
\Delta_L U(x) = 0 \quad in \quad \Omega, \quad U(x) = G(x) \quad on \quad \Gamma,
$$

which is equal to

$$
U(x) = G_0 \mathcal{P}_0 + \sum_{m,q=1}^\infty G_{mq} \hat{\mathcal{P}}_{mq}(x),
$$

where

$$
G_0 = \int_{\hat{L}_2(0,1)} G(x) \mu_w(dx), \quad G_{mq} = \int_{\hat{L}_2(0,1)} G(x) \mathcal{P}_{mq} \mu_w(dx),
$$

$$\hat{H}_{lv,q}(x) = \frac{1}{l!} \int\limits_{\hat{L}_2(0,1)} \prod_{k=1}^{l} \int_0^1 [x(t_k) + 2\sqrt{T(x)}\,Y(t_k)]^2 dt_k$$

$$\times \int_0^1 \cdots \int_0^1 s(t_{l+1}, \ldots, t_{l+q}) \prod_{j=l+1}^{l+q} [x(t_j) + 2\sqrt{T(x)}\,Y(t_j)]dt_j \mu_w(dy),$$

$$H_{lv,q}(x) = \frac{1}{l!} \prod_{k=1}^{l} \int_0^1 x^2(t_k)dt_k \int_0^1 \cdots \int_0^1 s(t_{l+1}, \ldots, t_{l+q}) \prod_{j=l+1}^{l+q} x(t_j)dt_j,$$

$$2l + v = m,$$

are the forms according to which the polynomials $\mathcal{P}_{mq}(x)$ *are constructed (see (2.4))*,

$$T(x) = \frac{1 - Q(x)}{\Delta_L Q(x)}, \quad Y(t_k) = \frac{y(\sigma_{t_k}) - y(\sigma_{t_{k-1}})}{\sqrt{t_k}}, \quad \sigma_{t_k} = \sum_{i=1}^{k} t_i; \quad \sigma_{t_0} = 0.$$

Proof. The forms $H_{lv,q}(x)$, and, therefore, the polynomials $\mathcal{P}_{mq}(x)$, are Gâteaux functionals satisfying the conditions of Theorem 5.8. Therefore, $\Delta_L \hat{\mathcal{P}}_{mq}(x) = 0$ in Ω, $\hat{\mathcal{P}}_{mq}\big|_{\Gamma} = \mathcal{P}_{mq}(x)$.

By Lemma 5.2 we have $\inf\limits_{x \in \Gamma} \hat{\mathcal{P}}_{mq}(x) = \inf\limits_{x \in \hat{\Omega}} \hat{\mathcal{P}}_{mq}(x) \le \hat{\mathcal{P}}_{mq}(x) \le \sup\limits_{x \in \hat{\Omega}} \hat{\mathcal{P}}_{mq}(x) = \sup\limits_{x \in \Gamma} \hat{\mathcal{P}}_{mq}(x)$ for a functional $\hat{\mathcal{P}}_{mq}(x)$ harmonic in Ω and continuous in $\bar{\Omega}$. Since $\hat{\mathcal{P}}_{mq}\big|_{\Gamma} = \mathcal{P}_{mq}(x)$, we have $|\hat{\mathcal{P}}_{mq}(x)| \le \sup\limits_{x \in \Gamma} |\mathcal{P}_{mq}(x)|$. By the absolute and uniform convergence on $\hat{\Omega}$ of the series $G_0 + \sum_{m,q}^{\infty} G_{mq} \mathcal{P}_{mq}(x)$, the series $G_0 + \sum_{m,q}^{\infty} G_{mq} \hat{\mathcal{P}}_{mq}(x)$ converges uniformly on Ω. The series $\sum_{m,q}^{\infty} G_{mq} d\hat{\mathcal{P}}_{mq}(x;h)$, $\sum_{m,q}^{\infty} G_{mq} d^2\hat{\mathcal{P}}_{mq}(x;h)$, $h \in L_2(0, 1)$, and $\sum_{m,q}^{\infty} G_{mq} \Delta_L \hat{\mathcal{P}}_{mq}(x)$ converge uniformly on Ω as well.

This along with the equality $\Delta_L \hat{\mathcal{P}}_{mq}(x) = 0$ in Ω, $\hat{\mathcal{P}}_{mq}(x)\big|_{\Gamma} = \mathcal{P}_{mq}(x)$, $m, q = 1, 2, \ldots$ means that $\Delta_L U(x) = 0$ in Ω $U\big|_{\Gamma} = G(x)$, and that $U(x)$ lies in the uniqueness class indicated in Theorem 5.4. $\square$

Theorem 5.11 *Let* $\Phi(x)$ *be a functional defined and continuous on* $L_2(0, 1)$, *and which admits the representation as the series in the polynomials* $\mathcal{P}_0$, $\mathcal{P}_{mq}(x)$, $m, q = 1, 2, \ldots$, *absolutely and uniformly converging on* $\bar{\Omega}$. *Then under the conditions of Theorem 5.10 there exists the unique solution of the problem*

$$\Delta_L U(x) = \Phi(x) \quad in \quad \Omega, \quad U(x) = G(x) \quad on \quad \Gamma,$$

which is equal to

$$U(x) = G_0 \mathcal{P}_0 + \sum_{m,q=1}^{\infty} G_{mq} \hat{\mathcal{P}}_{mq}(x) - \left(\Phi_0 \mathcal{P}_0 + \sum_{m,q=1}^{\infty} \Phi_{mq} \check{\mathcal{P}}_{mq}(x) \right),$$

where

$$\Phi_0 = \int_{\hat{L}_0(0,1)} \Phi(x) \mu_w(dx), \quad \Phi_{mq}(x) = \int_{\hat{L}_0(0,1)} \Phi(x) \mathcal{P}_{mq}(x) \mu_w(dx),$$

$$\check{H}_{lv,q}(x) = \frac{1}{l!} \int_0^{T(x)} \int_{\hat{L}_0(0,1)} \prod_{k=1}^{l} \int_0^1 [x(t_k) + 2\sqrt{T(x) - \xi}\, Y(t_k)]^2 dt_k$$

$$\times \int_0^1 \cdots \int_0^1 s(t_{l+1}), \ldots, s(t_{l+q}) \prod_{j=l+1}^{l+q} [x(t_j)$$

$$+ 2\sqrt{T(x) - \xi}\, Y(t_j)] dt_j \mu_w(dy) d\xi, \quad 2l + v = m.$$

Proof. Given the solution of the Dirichlet problem for the Lévy–Laplace equation (Theorem 5.10), it is sufficient to show that the solution of the Dirichlet problem with homogeneous boundary conditions for the Lévy–Poisson equation $\Delta_L V(x) = \Phi(x)$ in Ω, $V(x) = 0$ on Γ is equal to

$$V(x) = -\left(\Phi_0 \mathcal{P}_0 + \sum_{m,q=1}^{\infty} \Phi_{mq} \check{\mathcal{P}}_{mq}(x) \right).$$

The forms $H_{lv,q(x)}$, and, therefore the polynomials $\check{\mathcal{P}}_{mq}(x)$ as well, are Gâteaux functionals which satisfy the conditions of Theorem 5.9. Therefore $\Delta_L \check{\mathcal{P}}_{mq}(x) = -\mathcal{P}_{mq}(x)$ in Ω, $\check{\mathcal{P}}_{mq}\big|_{\Gamma} = 0$.

The final part of the proof does not differ from that given in Theorem 5.10. $\qquad \square$

5.2 The Dirichlet problem for the Lévy–Schrödinger stationary equation

We now consider the Dirichlet problem for the Lévy–Schrödinger equation

$$\Delta_L U(x) + P(x)U(x) = 0, \tag{5.10}$$

where $U(x)$ is a function to be determined and $P(x)$ is a given function on H.

Theorem 5.12 *Let $G(x)$ be a function defined and continuous in a bounded domain $\Omega \cup \Gamma$ of the space H.*

If $V(x)$ is a solution of the equation $\Delta_L V(x) = -P(x)$, then the solution of (5.10) is equal to

$$U(x) = \Psi(x)e^{V(x)},$$

where $\Psi(x)$ is an arbitrary harmonic function on H.

The solution of the Dirichlet problem

$$\Delta_L U(x) + P(x)U(x) = 0 \quad in \quad \Omega, \quad U(x) = G(x) \quad on \quad \Gamma$$

has the form

$$U(x) = \Psi(x)e^{V(x)},$$

where $V(x)$ is the solution of the Dirichlet problem for the Lévy–Poisson equation

$$\Delta_L V(x) = -P(x) \quad in \quad \Omega, \quad V\big|_{\Gamma} = 0,$$

and $\Psi(x)$ is the solution of the Dirichlet problem for the Lévy–Laplace equation

$$\Delta_L \Psi(x) = 0 \quad in \quad \Omega, \quad \Psi\big|_{\Gamma} = G(x).$$

The solution exists and is unique in the same functional classes where there exists the unique solution of the Dirichlet problem both for the Lévy–Laplace equation and the Lévy–Poisson equation.

Proof. By formula (1.4) for $m = 1$, we derive

$$\Delta_L \left[\ln \left| \frac{U(x)}{\Psi(x)} \right| \right] = \frac{\Delta_L U(x)}{U(x)}$$

and

$$\Delta_L V(x) = -P(x),$$

where $V(x) = \ln \left| \frac{U(x)}{\Psi(x)} \right|$, for $U(x) \neq 0$ and for arbitrary harmonic function $\Psi(x) \neq 0$.

Hence, $U(x) = \Psi(x)e^{V(x)}$.

The conditions $V\big|_{\Gamma} = 0$, and $\Psi\big|_{\Gamma} = G(x)$ ensure $V\big|_{\Gamma} = G(x)$.

The final statement of the theorem is clear. $\qquad \square$

Note that, in contrast with the finite-dimensional case, the condition $P(x) \leq 0$ is not required for the uniqueness of the solution of equation (5.10). This, in particular, follows from the uniqueness theorems for the solutions of the Lévy–Laplace and Lévy–Poisson equations given in Section 5.1.

5.3 The Riquier problem for the equation with iterated Lévy Laplacians

The Riquier problem for the equation with iterated Lévy Laplacians

$$F(x, U(x), \Delta_L U(x), \dots, \Delta_L^r U(x)) = \Phi(x),$$

where the function $F(x, u_1, \dots, u_r)$ is defined on $H \times \mathbb{R}^r$, consists in finding a function $U(x)$ satisfying this equation in a bounded domain $\Omega \subset H$ and such that $U(x), \Delta_L U(x), \dots, \Delta_L^{r-1} U(x)$, are equal to given functions $G_0(x), G_1(x), \dots, G_{r-1}(x)$ on the boundary Γ of this domain.

It is especially simple to solve the Riquier problem for the polyharmonic equation $\Delta_L^r U(x) = 0$.

Given functions $G_0(x), G_1(x), \dots, G_{r-1}(x)$ defined and continuous in a bounded domain $\Omega \cup \Gamma$ of the space H, one can reduce the Riquier problem

$$\Delta_L^r U(x) = 0 \quad \text{in} \quad \Omega, \quad U(x) = G_0(x),$$
$$\Delta_L U(x) = G_1(x), \dots, \Delta_L^{r-1} U(x) = G_{r-1}(x) \quad \text{on} \quad \Gamma$$

to the Dirichlet problem for the Lévy–Laplace equation and to $(r-1)$ Dirichlet problems for the Lévy–Poisson equation.

Indeed, the Riquier problem for the polyharmonic equation can be reduced to the following system:

$$\Delta_L U_{r-1}(x) = 0 \quad \text{in} \quad \Omega, \quad U_{r-1}\big|_\Gamma = G_{r-1}(x),$$
$$\Delta_L U_{r-2}(x) = U_{r-1}(x) \quad \text{in} \quad \Omega, \quad U_{r-2}\big|_\Gamma = G_{r-2}(x), \dots,$$
$$\Delta_L U_1(x) = U_2(x) \quad \text{in} \quad \Omega, \quad U_1\big|_\Gamma = G_1(x),$$
$$\Delta_L U(x) = U_1(x) \quad \text{in} \quad \Omega, \quad U\big|_\Gamma = G_0(x).$$

It is clear that there exists a unique solution to the Riquier problem for the polyharmonic equation in the same functional classes in which there exists a unique solution of the Dirichlet problem both to the Lévy–Laplace and Lévy–Poisson equations.

Consider now the linear equation for iterated Lévy Laplacians with harmonic coefficients

$$\Delta_L^r U(x) + H_1(x)\Delta_L^{r-1} U(x) + \cdots + H_r(x)U(x) = 0, \tag{5.11}$$

where $H_j(x)$ are harmonic function on H, $j = 1, 2, \dots, r$.

The equation with constant coefficients

$$\Delta_L^r U(x) + a_1 \Delta_L^{r-1} U(x) + \cdots + a_r U(x) = 0$$

is a special case of equation (5.11).

Theorem 5.13 (E. M. Polishchuk) *Let* $G_0(x)$, $G_1(x)$, ..., $G_{r-1}(x)$ *be functions defined and continuous in a bounded domain* $\Omega \cup \Gamma$ *of the space* H.
If all roots $\Lambda_1(x)$, ..., $\Lambda_r(x)$ *of the equation*

$$\Lambda^r(x) + H_1(x)\Lambda^{r-1}(x) + \cdots + H_r(x) = 0 \tag{5.12}$$

are different, then the solution of equation (5.11) is equal to

$$U(x) = \sum_{j=1}^r \Psi_j(x) \exp\left\{\frac{1}{2}\Lambda_j(x)||x||_H^2\right\},$$

where $\Psi_j(x)$ *are arbitrary harmonic functions.*
If, in addition, for all $x \in \bar{\Omega}$, $\Lambda_j(x) \neq \Lambda_i(x)$, $j \neq i$, *then the solution of the Riquer problem*

$$\Delta_L^r U(x) + H_1(x)\Delta_L^{r-1} U(x) + \cdots + H_r(x)U(x) = 0 \quad in \quad \Omega,$$
$$U(x) = G_0(x), \ \Delta_L U(x) = G_1(x), \ldots, \Delta_L^{r-1} U(x) = G_{r-1}(x) \quad on \quad \Gamma$$

has the form

$$U(x) = \sum_{j=1}^r \Psi_j(x) \exp\left\{\frac{1}{2}\Lambda_j(x)||x||_H^2\right\}, \tag{5.13}$$

where $\Psi_j(x)$ *satisfy* r *Dirichlet problems for the Lévy–Laplace equation:*

$$\Delta_L \Psi_j(x) = 0 \quad in \quad \Omega, \quad \Psi_j(x)\big|_\Gamma = \Lambda_j(x), \quad j = 1, \ldots, r,$$

and $\Lambda_j(x)$ *solve the system of linear equations*

$$\sum_{j=1}^r \Psi_j(x)\Lambda_j^m(x) \exp\left\{\frac{1}{2}\Lambda_j(x)||x||_H^2\right\} = G_m(x), \ m = 0, \ldots, r - 1. \tag{5.14}$$

Proof. By (1.4), $\Lambda_j(x)$ are harmonic functions. Since the $\Psi_j(x)$ are harmonic functions as well, and $\Delta_L||x||_H^2 = 2$, then by (1.4),

$$\Delta_L^m U(x) = \sum_{j=1}^r \Psi_j(x)\Lambda_j^m(x) \exp\left\{\frac{1}{2}\Lambda_j(x)||x||_H^2\right\}, \quad m = 1, \ldots, r.$$

Substituting these expressions into (5.11), and taking into account (5.12), we obtain an identity.

The determinant of the system (5.14) is equal to

$$\exp\left\{\frac{1}{2}\sum_{j=1}^{r}\Lambda_j(x)||x||_H^2\right\}[\Lambda_1(x)-\Lambda_2(x)]\cdots[\Lambda_1(x)-\Lambda_r(x)]$$
$$\times[\Lambda_2(x)-\Lambda_3(x)]\cdots[\Lambda_2(x)-\Lambda_r(x)]\times\cdots\times[\Lambda_{r-1}(x)-\Lambda_r(x)].$$

It is not equal to zero, because, according to the theorem condition, we have $\Lambda_i(x)\neq\Lambda_j(x)$ for $x\in\bar{\Omega}$, $i\neq j$. Therefore, we have reduced the Riquier problem to r Dirichlet problems for Lévy–Laplace equations. $\qquad\square$

Note that solution (5.13) exists and is unique in the same functional classes where there exists a unique solution of the Dirichlet problem for the Lévy–Laplace equation.

In a similar way one can write the solution of the Riquier problem in the case of multiple roots of equation (5.12). However, one should take into account that if the root $\Lambda_l(x)$ has multiplicity k_l, $l=1,\dots,p$, then

$$U(x)=\sum_{l=1}^{p}\left[\Psi_{l1}(x)+\Psi_{l2}(x)\frac{||x||_H^2}{2}+\cdots\right.$$
$$\left.+\Psi_{lk_l}(x)\left(\frac{||x||_H^2}{2}\right)^{k_l-1}\right]\exp\left\{\frac{1}{2}\Lambda_l(x)||x||_H^2\right\},$$

where $\Psi_{lk_l}(x)$ are harmonic functions ($l=1,\dots,p$), $\sum_{l=1}^{p}k_l=r$.

5.4 The Cauchy problem for the heat equation

Consider the heat equation

$$\frac{\partial U(t,x)}{\partial t}=\Delta_L U(t,x),$$

where $U(t,x)$ is a function on $[0,T]\times H$.

It follows from the work of Polishchuk [118], that the Cauchy problem for the heat equation

$$\frac{\partial U(t,x)}{\partial t}=\Delta_L U(t,x),\quad U(0,x)=G(x)$$

corresponds to its dual problem, i.e., the Dirichlet problem for the Lévy–Laplace equation

$$\Delta_L U(x)=0\quad\text{in}\quad\Omega,\quad U=G(x)\quad\text{on}\quad\Gamma.$$

The part of the time variable t is played by the function

$$T(x) = \frac{1 - Q(x)}{\Delta_L Q(x)}; \quad T(x) > 0 \text{ at } x \in \Omega, \quad T(x) = 0 \text{ at } x \in \Gamma,$$

which is defined on the fundamental domain

$$\Omega \cup \Gamma = \{x \in H : 0 \le Q(x) \le 1, \quad \Delta_L Q(x) = \text{const.}\}.$$

Based on theorems from Section 5.1 we can formulate the following theorems.

For the class of functions of the form $\Phi(x) = \phi(x, (Ax, x)_H)$, where $A = \gamma_A E + S_A$, $\phi(x, \xi)$ is a function on $H \times \mathbb{R}^1$, such that $\phi(x, (Ax, x)_H) = \phi(Px, (Ax, x)_H)$, then P is a projection on an m-dimensional subspace:

Theorem 5.14 *Let* $G(x) = g(x, (Ax, x)_H)$, $A = \gamma_A E + S_A$. *Then*

$$U(x) = g(x, 2\gamma_A t + (Ax, x)_H)$$

is the solution of the problem

$$\frac{\partial U(t, x)}{\partial t} = \Delta_L U(t, x), \quad U(0, x) = G(x).$$

For the class of Gâteaux functionals:

Theorem 5.15 *Let*

$$G(x) = \int_0^1 \cdots \int_0^1 g(x(t_1), \ldots, x(t_N); t_1, \ldots, t_N) dt_1 \cdots dt_N,$$

be a functional from the Gâteaux class. Then

$$U(x) = \frac{1}{(4\pi t)^{N/2}} \int_0^1 \cdots \int_0^1 dt_1 \cdots dt_N$$

$$\times \int_{-\infty}^{\infty} \cdots \int_{-\infty}^{\infty} g(\zeta_1, \ldots, \zeta_N; t_1, \ldots, t_N) \exp - \sum_{k=1}^{N} \frac{(x(t_k) - \zeta_k)^2}{4t} d\zeta_1 \cdots d\zeta_N$$

is the unique solution of the problem

$$\frac{\partial U(t, x)}{\partial t} = \Delta_L U(t, x), \quad U(0, x) = G(x).$$

Theorem 5.16 *Let*

$$G(x) = \int_0^1 \cdots \int_0^1 g(x(t_1), \ldots, x(t_N); t_1, \ldots, t_N) dt_1 \cdots dt_N,$$

be a functional from the Gâteaux class. Let $g(\xi_1, \ldots, \xi_N; t_1, \ldots, t_N)$
be twice continuously differentiable in $\xi_1, \ldots, \xi_N$ *and, together with its
derivatives, integrable in* $\mathbb{R}^N$ *with respect to the measure having density*
$(1/(2\pi)^{N/2}) \exp(-\frac{1}{2} \sum_{i=1}^{N} \xi_i^2)$. *Then there exists a unique solution of the problem*

$$\frac{\partial U(t, x)}{\partial t} = \Delta_L U(t, x), \quad U(0, x) = G(x),$$

which is equal to

$$U(x) = \int_{\hat{L}_2(0,1)} \int_0^1 \cdots \int_0^1 g(x(t_1) + 2\sqrt{t}Y(t_1), \ldots, x(t_N)$$
$$+ 2\sqrt{t}Y(t_N); t_1, \ldots, t_N)dt_1 \cdots dt_N \mu_w(dy),$$

where

$$Y(t_k) = \frac{y(\sigma_{t_k}) - y(\sigma_{t_{k-1}})}{\sqrt{t_k}}; \quad \sigma_{t_k} = \sum_{i=1}^{k} t_i, \quad \sigma_{t_0} = 0.$$

For the class of functionals which admit the representation series in orthopolynomials:

Theorem 5.17 *Let the functional* $G(x)$ *be defined and continuous on* $L_2(0, 1)$
*and be represented in the form of an absolutely and uniformly converging series
in the polynomials* $\mathcal{P}_0, \mathcal{P}_{mq}(x)$, $m, q = 1, \ldots$. *Then there exists the unique
solution of the problem*

$$\frac{\partial U(t, x)}{\partial t} = \Delta_L U(t, x), \quad U(0, x) = G(x),$$

which is equal to

$$U(x) = G_0 \mathcal{P}_0 + \sum_{m,q=1}^{\infty} G_{mq} \hat{\mathcal{P}}_{mq}(x),$$

where

$$G_0 = \int_{\hat{L}_2(0,1)} G(x)\mu_w(dx), \quad G_{mq} = \int_{\hat{L}_2(0,1)} G(x)\mathcal{P}_{mq}\mu_w(dx),$$

$$\hat{H}_{lv,q}(x) = \frac{1}{l!} \int_{\hat{L}_2(0,1)} \prod_{k=1}^{l} \int_0^1 [x(t_k) + 2\sqrt{t}Y(t_k)]^2 dt_k$$

$$\times \int_0^1 \cdots \int_0^1 s(t_{l+1}, \ldots, t_{l+q}) \prod_{j=l+1}^{l+q} [x(t_j) + 2\sqrt{t}Y(t_j)]dt_j \mu_w(dy),$$

$$2l + v = m,$$

$$H_{lv,q}(x) = \frac{1}{l!} \prod_{k=1}^l \int_0^1 x^2(t_k)dt_k \int_0^1 \cdots \int_0^1 s(t_{l+1}, \ldots, t_{l+q}) \prod_{j=l+1}^{l+q} x(t_j)dt_j,$$

$$2l + v = m$$

are the forms according to which the polynomials $\mathcal{P}_{mq}(x)$ *are constructed (see* (2.4)), $Y(t_k) = (y(\sigma_{t_k}) - y(\sigma_{t_{k-1}}))/\sqrt{t_k};$ $\sigma_{t_k} = \sum_{i=1}^k t_i,$ $\sigma_{t_0} = 0.$

Proofs of Theorems 5.14, 5.15, 5.16, and 5.17 coincide with proofs of Theorems 5.3, 5.6, 5.8, and 5.10, respectively.

6

Quasilinear and nonlinear elliptic equations with Lévy Laplacians

Solutions of Dirichlet and Riquier problems for quasilinear and nonlinear elliptic equations are constructed in the same functional classes in which the solution of the Dirichlet problem for the Lévy–Laplace equation exists (the reduction of the problem). This allows us to cover various functional classes, since the Dirichlet problem for linear equations with Lévy Laplacians has been solved in numerous works for a wide variety of classes.

6.1 The Dirichlet problem for the equation $\Delta_L U(x) = f(U(x))$

It was shown by Lévy that a general solution of the equation solved with respect to the Lévy Laplacian (quasilinear)

$$\Delta_L U(x) = f(U(x)), \tag{6.1}$$

where $U(x)$ is a function on H, and $f(\xi)$ is a given continuous function of a single variable in the range of $\{U(x)\}$ in $\mathbb{R}^1$, is given by the formula

$$\varphi(U(x)) - \frac{1}{2}||x||_H^2 = \Psi(x), \tag{6.2}$$

where

$$\varphi(\xi) = \int \frac{d\xi}{f(\xi)};$$

$\Psi(x)$ is a harmonic function on H. The solution of the Dirichlet problem

$$\Delta_L U(x) = f(U(x)) \quad \text{in} \quad \Omega, \quad U\big|_\Gamma = G(x),$$

where Ω is a bounded domain in H with a boundary Γ and $G(x)$ is a given function, is reduced to the solution of the Dirichlet problem for the Lévy–Laplace equation

$$\Delta_L \Psi(x) = 0 \quad \text{in} \quad \Omega, \quad \Psi\Big|_\Gamma = \varphi(G(x)) - \frac{1}{2}\|x\|_H^2\Big|_\Gamma.$$

Indeed, from (6.2) by (1.4) for $m = 1$, we obtain

$$\varphi_\xi'(U(x))\Delta_L U(x) - \frac{1}{2}\Delta_L\|x\|_H^2 = \Delta_L \Psi(x).$$

But $\varphi_\xi'(\xi) = 1/f(\xi)$; by (1.2), $\Delta_L\|x\|_H^2 = 2$, and $\Delta_L \Psi(x) = 0$; therefore

$$\Delta_L U(x) = f(U(x)).$$

It is also clear that, since (6.2) is solved with respect to $\Psi(x)$, the Dirichlet problem for equation (6.1) is reduced to the Dirichlet problem for the Lévy–Laplace equation.

Example 6.1 We solve the Dirichlet problem in a unit ball in the space H for the equation

$$\Delta_L U(x) = U^2(x),$$
$$U\Big|_{\|x\|_H^2=1} = (T(x - x_0), x - x_0)_H,$$

where T is a positive compact operator in H, $x_0 \in H$, and $\|x_0\|_H > 1$.
For this equation,

$$f(\xi) = \xi^2, \quad \varphi(\xi) = -\frac{1}{\xi},$$

and, by (6.2), its solution has the form

$$U(x) = -\frac{1}{\frac{1}{2}\|x\|_H^2 + \Psi(x)}.$$

It allows us to find $\Psi(x)$ on the surface $\|x\|_H^2 = 1$. The solution of the Dirichlet problem in a ball of the radius 1 for the Lévy–Laplace equation

$$\Delta_L \Psi(x) = 0, \quad \Psi\Big|_{\|x\|_H^2=1} = -\frac{2 + (T(x - x_0), x - x_0)_H}{2(T(x - x_0), x - x_0)_H}$$

has the form (see Section 5.1)

$$\Psi(x) = -\frac{2 + (T(x - x_0), x - x_0)_H}{2(T(x - x_0), x - x_0)_H}.$$

Substituting $\Psi(x)$ into the solution of the equation, we obtain the solution of the problem:

$$U(x) = \frac{2(T(x - x_0), x - x_0)_H}{(1 - \|x\|_H^2)(T(x - x_0), x - x_0)_H + 2}.$$

6.2 The Dirichlet problem for the equation
$$f(U(x), \Delta_L U(x)) = F(x)$$

Consider the nonlinear equation which is not solved with respect to the Lévy Laplacian

$$f(U(x), \Delta_L U(x)) = F(x), \tag{6.3}$$

where $U(x)$ is a function on H, and $F(x)$ is a given function defined on H, $f(\xi, \zeta)$ is a given function defined on R^2.

Theorem 6.1 *Let $f(\xi, \zeta)$ be a twice continuously differentiable function of two variables in the range of $\{U(x), \Delta_L U(x)\}$ in $\mathbb{R}^2$, and the function $F(x)$ satisfy the conditions $\Delta_L F(x) = \gamma \neq 0$, where γ is a constant. Then the solution of equation (6.3) (written implicitly) has the form*

$$f(U(x), \omega(U(x), \Psi(x))) - F(x) = 0, \tag{6.4}$$

where $\zeta = \omega(\xi, c)$ is a solution of the (nonlinear) ordinary differential equation

$$f'_\zeta(\xi, \zeta)\frac{d\zeta}{d\xi} + f'_\xi(\xi, \zeta) = \frac{\gamma}{\zeta}, \tag{6.5}$$

and $\Psi(x)$ is an arbitrary harmonic function on H.

If (6.4) is solvable with respect to $\Psi(x)$,

$$\Psi(x) = \varphi(U(x), F(x))$$

(here $\varphi(\xi, \beta)$ is some function on $\mathbb{R}^2$), then the Dirichlet problem

$$f(U(x), \Delta_L U(x)) = F(x) \quad in \quad \Omega, \quad U\Big|_\Gamma = G(x),$$

is reduced to the Dirichlet problem for the Lévy–Laplace equation

$$\Delta_L \Psi(x) = 0 \quad in \quad \Omega, \quad \Psi\Big|_\Gamma = \varphi(G(x), F(x)).$$

If, in addition, (6.4) is uniquely solvable with respect to $\Psi(x)$, and if the Dirichlet problem for the Lévy–Laplace equation has a unique solution in some functional class, then the solution of the Dirichlet problem for equation (6.3) is unique in this class.

Proof. Applying (1.4) twice with $m = 2$, we obtain from (6.4) that

$$f'_\xi(U, \omega(U, \Psi))\Delta_L U(x) + f'_\zeta(U, \omega(U, \Psi))$$

$$\times \left[\frac{\partial \omega}{\partial \xi}\Delta_L U(x) + \frac{\partial \omega}{\partial c}\Delta_L \Psi(x)\right] - \Delta_L F(x) = 0.$$

Since $\Delta_L \Psi(x) = 0$, $\Delta_L F(x) = \gamma$, we have

$$\left[f'_\xi(U, \omega(U, \Psi)) + f'_\zeta(U, \omega(U, \Psi)) \frac{\partial \omega}{\partial \xi} \right] \Delta_L U(x) = \gamma.$$

But, according to the condition of the theorem, $\omega(\xi, c)$ satisfies the equation (6.5), i.e.,

$$f'_\xi(U, \omega) + f'_\zeta(U, \omega) \frac{d\omega}{d\xi} = \frac{\gamma}{\omega};$$

therefore, $\frac{\gamma}{\omega} \Delta_L U = \gamma$. Hence

$$\Delta_L U = \omega(U, \Psi).$$

Substituting this into (6.3) and taking into account (6.4), we obtain the identity.
The final conclusions of Theorem 6.1 are obvious. $\qquad\square$

Example 6.2 We solve the Dirichlet problem in a unit ball of the space l_2 for the equation

$$||x||^2_{l_2}[\Delta_L U(x)]^2 - 4U(x)\Delta_L U(x) + ||x||^2_{l_2} = 0,$$

$$U\Big|_{||x||^2_{l_2}=1} = \frac{1}{2} \cosh \sum_{k=1}^{\infty} x_k^m, \quad m \geq 3.$$

Rewrite it in the form

$$\frac{4U(x)\Delta_L U(x)}{[\Delta_L U(x)]^2 + 1} = ||x||^2_{l_2}.$$

For this equation we have

$$f(\xi, \zeta) = \frac{4\xi\zeta}{(\xi^2 + 1)}, \quad \gamma = 2,$$

and (6.5) takes the form

$$\frac{2\xi\zeta}{(\xi^2 + 1)} \frac{d\zeta}{d\xi} = 1.$$

Its solution is $\zeta = \pm\sqrt{2\xi/c - 1}$.
Substituting the expression $\omega(U, \Psi) = \pm\sqrt{(2U(x)/\Psi(x)) - 1}$, into (6.4), we obtain the solution of the equation:

$$U(x) = \frac{\frac{1}{4}||x||^4_{l_2} + \Psi^2(x)}{2\Psi(x)}.$$

It allows us to find $\Psi(x)$ on the surface $||x||^2_{l_2} = 1$ which corresponds to the boundary condition problem. The solution of the Dirichlet problem in a unit ball for the Lévy–Laplace equation

$$\Delta_L \Psi(x) = 0, \quad \Psi\Big|_{\|x\|_{l_2}^2 = 1} = \frac{1}{2} \exp\left\{\pm \sum_{k=1}^{\infty} x_k^m\right\}$$

has the form (see Section 5.1)

$$\Psi(x) = \frac{1}{2} \exp\left\{\pm \sum_{k=1}^{\infty} x_k^m\right\}.$$

Substituting $\Psi(x)$ into the solution of the equation, we obtain the solution of the problem:

$$U(x) = \frac{1}{4}\left[\|x\|_{l_2}^4 \exp\left\{\mp \sum_{k=1}^{\infty} x_k^m\right\} + \exp\left\{\pm \sum_{k=1}^{\infty} x_k^m\right\}\right].$$

6.3 The Riquier problem for the equation $\Delta_L^2 U(x) = f(U(x))$

Consider the equation solved with respect to the iterated Lévy Laplacian (quasilinear)

$$\Delta_L^2 U(x) = f(U(x)), \tag{6.6}$$

where $U(x)$ is a function on H, and $f(\xi)$ is a given function on $\mathbb{R}^1$.

Theorem 6.2 *Let $f(\xi)$ be a continuous function of a single variable in the range of $\{U(x)\}$ in $\mathbb{R}^1$. Then the solution of* (6.6) *(written implicitly) has the form*

$$\Phi_0(U(x), \Psi_0(x), \Psi_1(x)) = \frac{1}{2}\|x\|_H^2, \tag{6.7}$$

where

$$\Phi_0(\xi, c_0, c_1) = \int \frac{d\xi}{\Phi_1(\xi, c_1)} + c_0,$$

$$\Phi_1(\xi, c_1) = \pm\sqrt{2 \int f(\xi)\, d\xi + c_1};$$

$\Psi_0(x), \Psi_1(x)$ *are arbitrary harmonic functions on H. In addition,*

$$\Delta_L U(x) = \Phi_1(U(x), \Psi_1(x)). \tag{6.8}$$

If (6.7), (6.8) *are solvable with respect to* $\Psi_0(x), \Psi_1(x)$:

$$\Psi_0(x) = \varphi_0\big(\|x\|_H^2, U(x), \Delta_L U(x)\big),$$

$$\Psi_1(x) = \varphi_1\big(U(x), \Delta_L U(x)\big)$$

(here $\varphi_0(\alpha, \xi, \zeta)$ is a function on $\mathbb{R}^3$, $\varphi_1(\xi, \zeta)$ is a function on $\mathbb{R}^2$), then the Riquier problem

$$\Delta_L^2 U(x) = f(U(x)) \quad \text{in} \quad \Omega, \quad U\big|_\Gamma = G_0(x), \Delta_L U(x)\big|_\Gamma = G_1(x),$$

where $G_0(x)$, $G_1(x)$ are some given functions, is reduced to a couple of Dirichlet problems for the Lévy–Laplace equations: the solution to the Riquier problem has the form of (6.7), where $\Psi_0(x)$, $\Psi_1(x)$ are solutions of the problems

$$\Delta_L \Psi_0(x) = 0 \quad \text{in} \quad \Omega, \quad \Psi_0\big|_\Gamma = \varphi_0\Big(||x||_H^2\big|_\Gamma, G_0(x), G_1(x)\Big),$$

$$\Delta_L \Psi_1(x) = 0 \quad \text{in} \quad \Omega, \quad \Psi_1\big|_\Gamma = \varphi_1(G_0(x), G_1(x)).$$

If, in addition, (6.7), (6.8) are uniquely solvable with respect to $\Psi_0(x)$, $\Psi_1(x)$, and if the Dirichlet problem for the Lévy–Laplace equation has a unique solution in some functional class, then the solution of the Riquer problem for equation (6.6) is unique in the same class.

Proof. From (6.7) by (1.4) for $m = 3$, we obtain

$$\frac{\partial \Phi_0(U, \Psi_0, \Psi_1)}{\partial \xi} \Delta_L U(x) + \frac{\partial \Phi_0(U, \Psi_0, \Psi_1)}{\partial c_0} \Delta_L \Psi_0(x)$$

$$+ \frac{\partial \Phi_0(U, \Psi_0, \Psi_1)}{\partial c_1} \Delta_L \Psi_1(x) = \frac{1}{2} \Delta_L ||x||_H^2.$$

In so far as

$$\frac{\partial \Phi_0(\xi, c_0, c_1)}{\partial \xi} = \frac{1}{\Phi_1(\xi, c_1)}, \quad \Delta_L \Psi_0(x) = \Delta_L \Psi_1(x) = 0,$$

and $\Delta_L ||x||_H^2 = 2$, we have

$$\Delta_L U(x) = \Phi_1(U(x), \Psi_1(x))$$

(using formula (6.8)). By (1.4) for $m = 2$, we obtain

$$\Delta_L^2 U(x) = \frac{\partial \Phi_1(U, \Psi_1)}{\partial \xi} \Delta_L U(x) + \frac{\partial \Phi_1(U, \Psi_1)}{\partial c_1} \Delta_L \Psi_1(x).$$

In so far as

$$\frac{\partial \Phi_1(\xi, c_1)}{\partial \xi} = \frac{f(\xi)}{\pm\sqrt{2 \int f(\xi) d\xi + c_1}} = \frac{f(\xi)}{\Phi_1(\xi, c_1)},$$

and $\Delta_L \Psi_1(x) = 0$, we have

$$\Delta_L^2 U(x) = \frac{f(U(x))}{\Phi_1(U(x), \Psi_1(x))} \Delta_L U(x).$$

But $\Delta_L U(x) = \Phi_1(U(x), \Psi_1(x))$; therefore

$$\Delta_L^2 U(x) = f(U(x)).$$

The solution $U(x)$ (given implicitly by (6.7)) satisfies equation (6.6) in Ω if $\Delta_L \Psi_0(x) = 0$ in Ω, and $\Delta_L \Psi_1(x) = 0$ in Ω. By the assumptions of the theorem we get $\Psi_0\big|_\Gamma = \varphi_0(\|x\|_H^2\big|_\Gamma, G_0(x), G_1(x))$, $\Psi_1\big|_\Gamma = \varphi_1(G_0(x), G_1(x))$.

Hence the solution to the Riquier problem has the form of (6.7), where $\Psi_0(x)$, $\Psi_1(x)$ are solutions of the Dirichlet problems

$$\Delta_L \Psi_0(x) = 0 \quad \text{in} \quad \Omega, \quad \Psi_0\big|_\Gamma = \varphi_0\left(\|x\|_H^2\big|_\Gamma, G_0(x), G_1(x)\right),$$

$$\Delta_L \Psi_1(x) = 0 \quad \text{in} \quad \Omega, \quad \Psi_1\big|_\Gamma = \varphi_1(G_0(x), G_1(x)).$$

The final conclusion of Theorem 6.2 are clear.   □

Example 6.3 We solve the Riquier problem in a unit ball of $L_2(0, 1)$ for the equation

$$\Delta_L^2 U(x) = e^{U(x)},$$

$$U\big|_{\|x\|_{L_2(0,1)}^2=1} = \int_0^1 \cos x(s)\, ds,$$

$$\Delta_L U(x)\big|_{\|x\|_{L_2(0,1)}^2=1} = \sqrt{3}\exp\left\{\frac{1}{2}\int_0^1 \cos x(s)\, ds\right\}.$$

For this equation, we have

$$f(\xi) = e^\xi,$$

therefore

$$\Phi_1(\xi, c_1) = \pm\sqrt{2e^\xi + c_1},$$

$$\Phi_0(\xi, c_0, c_1) = \pm\frac{1}{\sqrt{c_1}}\ln\frac{\sqrt{2e^\xi + c_1} - \sqrt{c_1}}{\sqrt{2e^\xi + c_1} + \sqrt{c_1}} + c_0, \quad c_1 > 0.$$

According to (6.7) we obtain the solution of the equation

$$U(x) = \ln\left\{\frac{1}{2}\Psi_1(x)sh^{-2}\left[\frac{1}{2}\sqrt{\Psi_1(x)}\left(\frac{1}{2}\|x\|_{L_2(0,1)}^2 - \Psi_0(x)\right)\right]\right\}.$$

In addition,

$$\Delta_L U(x) = \pm\sqrt{2e^{U(x)} - \Psi_1(x)}.$$

It allows us to find $\Psi_0(x)$ and $\Psi_1(x)$ on the surface $\|x\|_{L_2(0,1)}^2 = 1$ corresponding to the boundary conditions. The solutions of the Dirichlet problems in a unit ball of $L_2(0, 1)$ for the Lévy–Laplace equation

$$\Delta_L \Psi_0(x) = 0, \quad \Psi_0\Big|_{\|x\|^2_{L_2(0,1)}=1} = \frac{1}{2} - \exp\left\{-\frac{1}{2}\int_0^1 \cos x(s)\,ds\right\} \ln(2 \pm \sqrt{3})$$

and

$$\Delta_L \Psi_1(x) = 0, \quad \Psi_1\Big|_{\|x\|^2_{L_2(0,1)}=1} = \exp\left\{\int_0^1 \cos x(s)\,ds\right\}$$

have the form (see Section 5.1)

$$\Psi_0(x) = \frac{1}{2} - \exp\left\{-\frac{1}{2}\exp\left\{-\frac{1}{2}\left(1 - \int_0^1 x^2(s)\,ds\right)\right\}\right.$$

$$\times \int_0^1 \cos x(s)\,ds\Big\} \ln(2 \pm \sqrt{3}),$$

$$\Psi_1(x) = \exp\left\{\exp\left\{-\frac{1}{2}\left(1 - \int_0^1 x^2(s)\,ds\right)\right\}\int_0^1 \cos x(s)\,ds\right\}.$$

Substituting $\Psi_0(x)$ and $\Psi_1(x)$ into the solution of the equation, we obtain the solution of the problem:

$$U(x) = \ln\left(\frac{1}{2}\exp\left\{\exp\left\{-\frac{1}{2}\left(1 - \int_0^1 x^2(s)\,ds\right)\right\}\int_0^1 \cos x(s)\,ds\right\}\right.$$

$$\times \sinh^{-2}\left[\frac{1}{4}\exp\left\{\frac{1}{2}\exp\left\{-\frac{1}{2}\left(1 - \int_0^1 x^2(s)\,ds\right)\right\}\int_0^1 \cos x(s))\,ds\right\}\right.$$

$$\times \left(\|x\|^2_{L_2(0,1)} - 1\right) + \frac{1}{2}\ln\left(2 \pm \sqrt{3}\right)\Big]\Big).$$

6.4 The Riquier problem for the equation $f(U(x), \Delta_L^2 U(x)) = \Delta_L U(x)$

Consider the nonlinear equation not solved with respect to the iterated Lévy Laplacian, but solved with respect to the Lévy Laplacian

$$f(U(x), \Delta_L^2 U(x)) = \Delta_L U(x), \tag{6.9}$$

where $U(x)$ is a function on H, and $f(\xi, \zeta)$ is a given function on $\mathbb{R}^2$.

Theorem 6.3 *Let $f(\xi, \zeta)$ be a twice continuously differentiable function of two variables in the range of $\{U(x), \Delta_L^2 U(x)\}$ in $\mathbb{R}^2$. If $\zeta = \omega(\xi, c_0)$ is the solution of the (nonlinear) ordinary differential equation*

$$f'_\zeta(\xi, \zeta)\frac{d\zeta}{d\xi} + f'_\xi(\xi, \zeta) = \frac{\zeta}{f(\xi, \zeta)}, \tag{6.10}$$

then the solution of (6.9) (written implicitly) has the form

$$\Phi(U(x), \Psi_0(x), \Psi_1(x)) = \frac{1}{2}||x||_H^2, \tag{6.11}$$

where

$$\Phi(\xi, c_0, c_1) = \int \frac{d\xi}{f(\xi, \omega(\xi, c_0))} + c_1;$$

$\Psi_0(x), \Psi_1(x)$ *are the arbitrary harmonic functions on H. In addition,*

$$\Delta_L U(x) = f(U(x), \omega(U(x), \Psi_0(x))). \tag{6.12}$$

If (6.11) and (6.12) are solvable with respect to $\Psi_0(x), \Psi_1(x)$:

$$\Psi_0(x) = \varphi_0(U(x), \Delta_L U(x)),$$
$$\Psi_1(x) = \varphi_1(||x||_H^2, U(x), \Delta_L U(x))$$

(here $\varphi_0(\xi, \zeta)$ is a function on $\mathbb{R}^2$, $\varphi_1(\alpha, \xi, \zeta)$ is a function on $\mathbb{R}^3$), then the Riquier problem

$$f(U(x), \Delta_L^2 U(x)) = \Delta_L U(x) \quad in \quad \Omega,$$
$$U\Big|_\Gamma = G_0(x), \quad \Delta_L U(x)\Big|_\Gamma = G_1(x),$$

where $G_0(x), G_1(x)$ are some given functions, is reduced to a couple of Dirichlet problems for the Lévy–Laplace equations: the solution to the Riquier problem has the form (6.11), where $\Psi_0(x), \Psi_1(x)$ are solutions of the problems

$$\Delta_L \Psi_0(x) = 0 \quad in \quad \Omega, \quad \Psi_0\Big|_\Gamma = \varphi_0(G_0(x), G_1(x)),$$

$$\Delta_L \Psi_1(x) = 0 \quad in \quad \Omega, \quad \Psi_1\Big|_\Gamma = \varphi_1\left(||x||_H^2\Big|_\Gamma, G_0(x), G_1(x)\right).$$

If, in addition, (6.11), (6.12) are uniquely solvable with respect to $\Psi_0(x), \Psi_1(x)$ and if the Dirichlet problem for the Lévy–Laplace equation has a unique solution in some functional class, then the solution of the Riquer problem for equation (6.9) is unique in the same class.

Proof. From (6.11) by (1.4) for $m = 3$, we obtain

$$\Phi'_\xi(U, \Psi_0, \Psi_1)\Delta_L U(x) + \Phi'_{c_0}(U, \Psi_0, \Psi_1)\Delta_L \Psi_0(x)$$

$$+ \Phi'_{c_1}(U, \Psi_0, \Psi_1)\Delta_L \Psi_1(x) = \frac{1}{2}\Delta_L \|x\|_H^2.$$

In so far as

$$\Phi'_\xi(\xi, c_0, c_1) = \frac{1}{f(\xi, \omega(\xi, c_0))}, \quad \Delta_L \Psi_0(x) = \Delta_L \Psi_1(x) = 0,$$

and $\Delta_L \|x\|_H^2 = 2$, we have

$$\Delta_L U(x) = f(U(x), \omega(U(x), \Psi_0(x)))$$

(formula (6.12)).

Applying (1.4) for $m = 2$, twice we obtain

$$\Delta_L^2 U(x) = f'_\xi(U, \omega(U, \Psi_0))\Delta_L U(x)$$

$$+ f'_\zeta(U, \omega(U, \Psi_0))\Delta_L \omega(U, \Psi_0)$$

$$= f'_\xi(U, \omega(U, \Psi_0))\Delta_L U(x)$$

$$+ f'_\zeta(U, \omega(U, \Psi_0))\left[\frac{\partial\omega(U, \Psi_0)}{\partial\xi}\Delta_L U(x) + \frac{\partial\omega(U, \Psi_0)}{\partial c_0}\Delta_L \Psi_0(x))\right].$$

In so far as $\Delta_L U(x) = f(U(x), \omega(U(x), \Psi_0(x)))$, and $\Delta_L \Psi_0(x) = 0$, we have

$$\Delta_L^2 U(x) = \left[f'_\xi(U, \omega(U, \Psi_0)) + f'_\zeta(U, \omega(U, \Psi_0))\frac{\partial\omega(U, \Psi_0)}{\partial\xi}\right]f(U, \omega(U, \Psi_0)).$$

But, by virtue of the condition of the theorem, $\omega(\xi, c_0)$ satisfies (6.10), i.e.,

$$f'_\xi(\xi, \omega(\xi, c_0)) + f'_\zeta(\xi, \omega(\xi, c_0))\frac{\partial\omega(\xi, c_0)}{\partial\xi} = \frac{\omega(\xi, c_0)}{f(\xi, \omega(\xi, c_0))};$$

therefore

$$\Delta_L^2 U(x) = \omega(U(x), \Psi_0(x)). \tag{6.13}$$

Substituting (6.13) into (6.9) and taking into account (6.12), we obtain the identity.

The solution $U(x)$ (given implicitly by (6.11)) satisfies the equation (6.9) in Ω if $\Delta_L \Psi_0(x) = 0$ in Ω, $\Delta_L \Psi_1(x) = 0$ in Ω. By the assumptions of the theorem we get $\Psi_0\big|_\Gamma = \varphi_0(G_0(x), G_1(x))$, $\Psi_1\big|_\Gamma = \varphi_1\left(\|x\|_H^2\big|_\Gamma, G_0(x), G_1(x)\right)$.

Hence the solution to the Riquier problem has the form of (6.11), where $\Psi_0(x)$, $\Psi_1(x)$ are solutions of the Dirichlet problems

$$\Delta_L \Psi_0(x) = 0 \quad \text{in} \quad \Omega, \quad \Psi_0\big|_\Gamma = \varphi_0(G_0(x), G_1(x)),$$

$$\Delta_L \Psi_1(x) = 0 \quad \text{in} \quad \Omega, \quad \Psi_1\big|_\Gamma = \varphi_1\left(\|x\|_H^2\big|_\Gamma, G_0(x), G_1(x)\right).$$

The final conclusion of Theorem 6.3 is clear.   □

Example 6.4 Let us solve the equation

$$[\Delta_L^2 U(x)]^n - nU^{n+1}(x)\Delta^2 U(x) + nU^n(x)[\Delta_L U(x)]^2 = 0. \tag{6.14}$$

Rewrite this equation in the form

$$\left\{ U(x)\Delta_L^2 U(x) - \frac{[\Delta_L^2 U(x)]^n}{nU^n(x)} \right\}^{1/2} = \Delta_L U(x).$$

For this equation, we have

$$f(\xi, \zeta) = \sqrt{\xi\zeta - \frac{\zeta^n}{n\xi^n}},$$

and (6.10) takes the form

$$[\xi^{n+2} - \xi\zeta^{n-1}]\frac{d\zeta}{d\xi} - \xi^{n+1}\zeta + \zeta^n = 0.$$

Its solution is $\zeta = c_0\xi$. Therefore

$$\Phi(\xi, c_0, c_1) = \frac{1}{c_0} \text{arc cosh} \frac{\sqrt{n}\xi}{c_0^{(n-1)/2}} + c_1$$

According to (6.11) we obtain the solution of the equation (6.14):

$$U(x) = \sqrt{\frac{\Psi_0^{n-1}(x)}{n}} \cosh\left[\sqrt{\Psi_0(x)}\left(\frac{1}{2}\|x\|_H^2 - \Psi_1(x)\right)\right], \tag{6.15}$$

where $\Psi_0(x), \Psi_1(x)$ are arbitrary harmonic functions.

Example 6.5 Let us solve the Riquier problem in a unit ball of the space H for equation (6.14) with $n = 3$

$$[\Delta_L^2 U(x)]^3 - 3U^4(x)\Delta_L^2 U(x) + 3U^3(x)[\Delta_L U(x)]^2 = 0,$$

$$U\Big|_{\|x\|_H^2=1} = (T(x - x_0), x - x_0)_H,$$

$$\Delta_L U\Big|_{\|x\|_H^2=1} = \sqrt{\frac{2}{3}}(T(x - x_0), x - x_0)_H^{\frac{3}{2}},$$

where T is a positive compact operator in H, $x_0 \in H$, and $\|x_0\|_H > 1$.

According to (6.15) we have

$$U(x) = \frac{\Psi_0(x)}{\sqrt{3}} \cosh\left[\sqrt{\Psi_0(x)}\left(\frac{\|x\|_H^2}{2} - \Psi_1(x)\right)\right].$$

In addition,

$$\Delta_L U(x) = \sqrt{\Psi_0(x)U^2(x) - \frac{1}{3}\Psi_0^3(x)}.$$

It allows us to find $\Psi_0(x)$ and $\Psi_1(x)$ on the surface $\|x\|_H^2 = 1$ corresponding to the boundary conditions. The solutions of the Dirichlet problems in a unit ball of H for the Lévy–Laplace equation

$$\Delta_L \Psi_0(x) = 0, \quad \Psi_0\Big|_{\|x\|_H^2=1} = (T(x - x_0), x - x_0)_H,$$

$$\Delta_L \Psi_1(x) = 0, \quad \Psi_1\Big|_{\|x\|_H^2=1} = \frac{1}{2} - \frac{\text{Arch}\sqrt{3}}{(T(x - x_0), x - x_0)_H^{1/2}},$$

have the form (see Section 5.1)

$$\Psi_0(x) = (T(x - x_0), x - x_0)_H,$$

$$\Psi_1(x) = \frac{1}{2} - \frac{\text{arc cosh}\sqrt{3}}{(T(x - x_0), x - x_0)_H^{1/2}}.$$

Substituting the above expressions for $\Psi_0(x)$ and $\Psi_1(x)$ into the solution of the equation, we obtain the solution of the problem:

$$U(x) = (T(x - x_0), x - x_0)_H \left\{ \cosh\left[\frac{1}{2}(T(x - x_0), x - x_0)_H^{1/2}(\|x\|_H^2 - 1)\right] \right.$$

$$\left. + \sqrt{\frac{2}{3}} \sinh\left[\frac{1}{2}(T(x - x_0), x - x_0)_H^{1/2}(\|x\|_H^2 - 1)\right] \right\}.$$

6.5 The Riquier problem for the equation $f(U(x), \Delta_L U(x), \Delta_L^2 U(x)) = 0$

Consider the nonlinear equation

$$f\left(U(x), \Delta_L U(x), \Delta_L^2 U(x)\right) = 0, \tag{6.16}$$

where $U(x)$ is a function on H, and $f(\xi, \eta, \zeta)$ is a given function on $\mathbb{R}^3$.

Theorem 6.4 *Let* $f(\xi, \eta, \zeta)$ *be a twice continuously differentiable function of three variables in the range of* $\{U(x), \Delta_L U(x), \Delta_L^2 U(x)\}$ *in* $\mathbb{R}^3$. *Let in addition the equation*

$$f(\xi, \eta, \rho\eta) = 0 \tag{6.17}$$

have a solution $\eta = \phi(\xi, \rho)$ *(although the equation* $f(\xi, \eta, \zeta) = 0$, *generally speaking is not solvable with respect to* η). *Then the solution of* (6.16) *(written implicitly) has the form*

$$\Phi(U(x), \Psi_0(x), \Psi_1(x)) = \frac{1}{2}\|x\|_H^2, \tag{6.18}$$

where

$$\Phi(\xi, c_0, c_1) = \int \frac{d\xi}{\phi(\xi, \omega(\xi, c_0))} + c_1,$$

$\rho = \omega(\xi, c_0)$ *is the solution of the (nonlinear) ordinary differential equation*

$$\phi'_\rho(\xi, \rho)\frac{d\rho}{d\xi} + \phi'_\xi(\xi, \rho) = \rho; \tag{6.19}$$

$\Psi_0(x)$, $\Psi_1(x)$ *are the arbitrary harmonic functions on* H. *In addition,*

$$\Delta_L U(x) = \phi(U(x), \omega(U(x), \Psi_0(x))). \tag{6.20}$$

If (6.18), (6.20) *are solvable with respect to* $\Psi_0(x)$, $\Psi_1(x)$:

$$\Psi_0(x) = \varphi_0(U(x), \Delta_L U(x)),$$
$$\Psi_1(x) = \varphi_1(\|x\|_H^2, U(x), \Delta_L U(x))$$

(here $\varphi_0(\xi, \eta)$ *is a function on* $\mathbb{R}^2$, $\varphi_1(\alpha, \xi, \eta)$ *is a function on* $\mathbb{R}^3$*), then the Riquier problem*

$$f\big(U(x), \Delta_L U(x), \Delta_L^2 U(x)\big) = 0 \quad in \quad \Omega,$$
$$U\Big|_\Gamma = G_0(x), \quad \Delta_L U(x)\Big|_\Gamma = G_1(x),$$

where $G_0(x)$, $G_1(x)$ *are some given functions, is reduced to a couple of Dirichlet problems for the Lévy–Laplace equations: the solution to the Riquier problem has the form of* (6.18), *where* $\Psi_0(x)$, $\Psi_1(x)$ *are solutions of the problems*

$$\Delta_L \Psi_0(x) = 0 \quad in \quad \Omega, \quad \Psi_0\Big|_\Gamma = \varphi_0(G_0(x), G_1(x)),$$
$$\Delta_L \Psi_1(x) = 0 \quad in \quad \Omega, \quad \Psi_1\Big|_\Gamma = \varphi_1\Big(\|x\|_H^2\Big|_\Gamma, G_0(x), G_1(x)\Big).$$

If, in addition, (6.18) *and* (6.20) *are uniquely solvable with respect to* $\Psi_0(x)$, $\Psi_1(x)$, *the equation* (6.17) *has a unique non-trivial solution* η, *and if the Dirichlet problem for the Lévy–Laplace equation has a unique solution in some functional class, then the solution of the Riquer problem for equation* (6.16) *is unique in the same class.*

Proof. From (6.18) by (1.4) for $m = 3$, we obtain

$$\Phi'_\xi(U, \Psi_0, \Psi_1)\Delta_L U(x) + \Phi'_{c_0}(U, \Psi_0, \Psi_1)\Delta_L \Psi_0(x)$$
$$+ \Phi'_{c_1}(U, \Psi_0, \Psi_1)\Delta_L \Psi_1(x) = \frac{1}{2}\Delta_L\|x\|_H^2.$$

In so far as

$$\Phi'_\xi(\xi, c_0, c_1) = \frac{1}{\phi(\xi, \omega(\xi, c_0))}, \qquad \Delta_L \Psi_0(x) = \Delta_L \Psi_1(x) = 0,$$

and $\Delta_L ||x||_H^2 = 2$, we have

$$\Delta_L U(x) = \phi(U(x), \omega(U(x), \Psi_0(x)))$$

(formula (6.20)).

Applying (1.4) for $m = 2$ twice, we obtain

$$\Delta_L^2 U(x) = \phi'_\xi(U, \omega(U, \Psi_0))\Delta_L U(x)$$
$$+ \phi'_\rho(U, \omega(U, \Psi_0))\Delta_L \omega(U, \Psi_0) = \phi'_\xi(U, \omega(U, \Psi_0))\Delta_L U(x)$$
$$+ \phi'_\rho(U, \omega(U, \Psi_0))\left[\frac{\partial\omega(U, \Psi_0)}{\partial\xi}\Delta_L U(x) + \frac{\partial\omega(U, \Psi_0)}{\partial c_0}\Delta_L \Psi_0(x))\right].$$

In so far as $\Delta_L U(x) = \phi(U(x), \omega(U(x), \Psi_0(x)))$, and $\Delta_L \Psi_0(x) = 0$, we have

$$\Delta_L^2 U(x) = \left[\phi'_\xi(U, \omega(U, \Psi_0) + \phi'_\rho(U, \omega(U, \Psi_0))\frac{\partial\omega(U, \Psi_0)}{\partial\xi}\right]\phi(U, \omega(U, \Psi_0)).$$

But, by virtue of the condition of the theorem, $\omega(\xi, c_0)$ satisfies (6.19), i.e.,

$$\phi'_\rho(\xi, \omega(\xi, c_0))\frac{d\omega(\xi, c_0)}{d\xi} + \phi'_\xi(\xi, \omega(\xi, c_0)) = \omega(\xi, c_0);$$

therefore

$$\Delta_L^2 U(x) = \omega(U(x), \Psi_0(x))\phi(U(x), \omega(U(x), \Psi_0(x))). \tag{6.21}$$

Substituting (6.20) and (6.21) into (6.16) we obtain the identity

$$f(U, \phi(U, \omega(U, \Psi_0)), \omega(U, \Psi_0)\phi(U, \omega(U, \Psi_0))) = 0,$$

since by the condition of the theorem the equality $\eta = \phi(\xi, \rho)$ follows from the equality $f(\xi, \eta, \rho\eta) = 0$.

The solution $U(x)$ (given implicitly by (6.18)) satisfies the equation (6.16) in Ω if $\Delta_L \Psi_0(x) = 0$ in Ω, and $\Delta_L \Psi_1(x) = 0$ in Ω. By the conditions of the theorem we get $\Psi_0\big|_\Gamma = \varphi_0(G_0(x), G_1(x))$, $\Psi_1\big|_\Gamma = \varphi_1\left(||x||_H^2\big|_\Gamma, G_0(x), G_1(x)\right)$.

Hence the solution to the Riquier problem has the form of (6.18), where $\Psi_0(x)$, $\Psi_1(x)$ are solutions of the Dirichlet problems

$$\Delta_L \Psi_0(x) = 0 \quad \text{in} \quad \Omega, \quad \Psi_0\big|_\Gamma = \varphi_0(G_0(x), G_1(x)),$$

$$\Delta_L \Psi_1(x) = 0 \quad \text{in} \quad \Omega, \quad \Psi_1\big|_\Gamma = \varphi_1\left(||x||_H^2\big|_\Gamma, G_0(x), G_1(x)\right).$$

The final conclusion of Theorem 6.4 is clear. $\qquad\qquad\qquad\qquad\qquad\square$

Example 6.6 Consider the Riquier problem in an ellipsoid $(Bx, x)_H < 1$ in the space H for the equation

$$\left[\Delta_L^2 U(x)\right]^2 - U(x)\Delta_L U(x)\Delta_L^2 U(x)$$
$$+ \frac{1}{2}U^2(x)[\Delta_L U(x)]^2 - [\Delta_L U(x)]^3 = 0, \tag{6.22}$$

$$U\Big|_{(Bx,x)_H=1} = 2e^{\|x\|_H^2}, \quad \Delta_L U(x)\Big|_{(Bx,x)_H=1} = e^{2\|x\|_H^2}. \tag{6.23}$$

Here $B = E + S$, where E is the identity operator and S is a self-adjoint compact operator in H.

For this equation we have

$$f(\xi, \eta, \zeta) = \zeta^2 - \xi\eta\zeta + \frac{1}{2}\xi^2\eta^2 - \eta^3 = 0$$

and (6.17) has the form

$$\rho^2\eta^2 - \rho\xi\eta^2 + \frac{1}{2}\xi^2\eta^2 - \eta^3 = 0.$$

This equation has the non-trivial solution

$$\eta = \phi(\xi, \rho) = \rho^2 - \rho\xi + \frac{\xi^2}{2},$$

and (6.19) takes the form

$$(2\rho - \xi)\frac{d\rho}{d\xi} + (\xi - \rho) = \rho.$$

Its solution is $\rho = \omega(\xi, c_0) = \xi + c_0$. Therefore

$$\phi(\xi, \omega(\xi, c_0)) = \frac{\xi^2}{2} + c_0\xi + c_0^2,$$

and

$$\Phi(\xi, c_0, c_1) = \int \frac{d\xi}{\xi^2/2 + c_0\xi + c_0^2} + c_1 = \frac{2}{c_0}\arctan\frac{\xi + c_0}{c_0} + c_1.$$

According to (6.18) we obtain the solution of the equation (6.22)

$$U(x) = \left\{\tan\left(\frac{1}{4}\left[\|x\|_H^2 - 2\Psi_1(x)\right]\Psi_0(x)\right) - 1\right\}\Psi_0(x). \tag{6.24}$$

In addition,

$$\Delta_L U(x) = \frac{1}{2}U^2(x) + \Psi_0(x)U(x) + \Psi_0^2(x).$$

It allows us to find $\Psi_0(x)$ and $\Psi_1(x)$ on the surface $(Bx, x)_H = 1$ corresponding to the boundary conditions. The solutions of the Dirichlet problems in a ellipsoid of H for the Lévy–Laplace equation

$$\Delta_L \Psi_0(x) = 0, \quad \Psi_0 \Big|_{(Bx,x)_H=1} = -e^{\|x\|_H^2},$$

$$\Delta_L \Psi_1(x) = 0, \quad \Psi_1 \Big|_{(Bx,x)_H=1} = \frac{1}{2}\Big[\|x\|_H^2 - \pi e^{-\|x\|_H^2}\Big]$$

have the form (see Section 5.1)

$$\Psi_0(x) = -e^{1-(Sx,x)_H}, \quad \Psi_1(x) = \frac{1}{2}[1 - (Sx, x)_H - \pi e^{-1+(Sx,x)_H}].$$

Substituting $\Psi_0(x)$ and $\Psi_1(x)$ into the solution (6.24) of the equation we obtain the solution of the problem (6.22), (6.23):

$$U(x) = \frac{2e^{1-(Sx,x)_H}}{1 + \tan\big(\frac{1}{4}e^{1-(Sx,x)_H}\big[1 - (Bx, x)_H\big]\big)}.$$

7

Nonlinear parabolic equations with
Lévy Laplacians

Solutions of the Cauchy problem for nonlinear parabolic equations are constructed in the same functional classes in which the solution of the Cauchy problem for the 'heat equation' exists (the reduction of the problem). This allows one to cover various functional classes, since the Cauchy problem for linear parabolic equations with Lévy Laplacians has been solved in numerous works for a wide variety of classes.

Each theorem of this chapter includes both a construction of the solution for the problem without initial data and a construction of the solution for the Cauchy problem. The solution to the Cauchy problem has the same form as the solution to the problem without initial data, but the arbitrary harmonic functions $\Psi(x)$ and $\Phi(x)$ in the representation of the solution to the problem without initial data are changed for the specially chosen functions $\Psi(t, x)$ and $\Phi(t, x)$.

7.1 The Cauchy problem for the equations
$(\partial U(t, x)/\partial t) = f(\Delta_L U(t, x))$ and
$(\partial U(t, x)/\partial t) = f(t, \Delta_L U(t, x))$

Consider the nonlinear equation

$$\frac{\partial U(t, x)}{\partial t} = f(\Delta_L U(t, x)), \tag{7.1}$$

where $U(t, x)$ is a function on $[0, T] \times H$, and $f(\zeta)$ is a given function of a single variable.

Theorem 7.1 *Let $f(\zeta)$ be a twice continuously differentiable function of a single variable in the range of $\{\Delta_L U(t, x)\}$ in $\mathbb{R}^1$.*

1. *Then the solution of equation (7.1) has the form*

$$U(t, x) = f(\Psi(x))t + \Psi(x)\frac{||x||_H^2}{2} + \Phi(x), \tag{7.2}$$

where $\Psi(x)$, $\Phi(x)$ are arbitrary harmonic functions on H.

2. *Let the equation*

$$f'_\zeta\left(\frac{\partial V(\tau, x)}{\partial \tau}\bigg|_{\tau=X-\frac{||x||_H^2}{2}}\right)t + \frac{||x||_H^2}{2} - X = 0, \tag{7.3}$$

where $V(\tau, x)$ is the solution of the Cauchy problem for the 'heat equation'

$$\frac{\partial V(\tau, x)}{\partial \tau} = \Delta_L V(\tau, x), \quad V(0, x) = U_0(x) \tag{7.4}$$

(and $(\partial^2 V(\tau, x)/\partial \tau^2) \neq 0$), is solvable with respect to $X = \chi(t, x)$ (and $\chi(0, x) = (1/2)||x||_H^2$). Then the solution of the Cauchy problem

$$\frac{\partial U(t, x)}{\partial t} = f(\Delta_L U(x)), \quad U(0, x) = U_0(x)$$

for the nonlinear equation (7.1) has the form

$$U(t, x) = f(\Psi(t, x))t + \Psi(t, x)\frac{||x||_H^2}{2} + \Phi(t, x), \tag{7.5}$$

where

$$\Psi(t, x) = \frac{\partial V(\tau, x)}{\partial \tau}\bigg|_{\tau=\chi(t,x)-\frac{||x||_H^2}{2}},$$

$$\Phi(t, x) = V\left(\chi(t, x) - \frac{||x||_H^2}{2}, x\right) - \Psi(t, x)\chi(t, x);$$

$V(\tau, x)$ is the solution of the Cauchy problem (7.4).

If, in addition, equation (7.3) is uniquely solvable with respect to $\chi(t, x)$ and if the Cauchy problem for the 'heat equation' has a unique solution in some functional class, then the solution of the Cauchy problem for equation (7.1) is unique in the same class.

Proof. 1. From (7.2) by (1.4), we obtain

$$\frac{\partial U(t, x)}{\partial t} = f(\Psi(x)),$$

$$\Delta_L U(t, x) = f'_\zeta(\Psi(x))t \Delta_L \Psi(x) + \frac{||x||_H^2}{2}\Delta_L \Psi(x)$$

$$+ \frac{1}{2}\Psi(x)\Delta_L ||x||_H^2 + \Delta_L \Phi(x) = \Psi(x),$$

since by harmonicity $\Delta_L \Psi(x) = \Delta_L \Phi(x) = 0$, and by (1.2), $\Delta_L \|x\|_H^2 = 2$.

Substituting these expressions into (7.1), we obtain the identity.

2. From (7.5) by (1.4), we obtain

$$
\frac{\partial U(t,x)}{\partial t} = f(\Psi(t,x)) + f_\zeta'(\Psi(t,x))t\frac{\partial \Psi(t,x)}{\partial t}
$$
$$
+ \frac{\|x\|_H^2}{2}\frac{\partial \Psi(t,x)}{\partial t} + \frac{\partial \Phi(t,x)}{\partial t}, \tag{7.6}
$$

$$
\Delta_L U(t,x) = f_\zeta'(\Psi(t,x))t\Delta_L \Psi(t,x)
$$
$$
+ \frac{\|x\|_H^2}{2}\Delta_L \Psi(t,x) + \Psi(t,x) + \Delta_L \Phi(t,x). \tag{7.7}
$$

By direct computation we find the expressions for $\partial \Phi(t,x)/\partial t$ and $\Delta_L \Phi(t,x)$:

$$
\frac{\partial \Phi(t,x)}{\partial t} = \frac{\partial V\left(\chi(t,x) - \frac{\|x\|_H^2}{2}, x\right)}{\partial t} - \frac{\partial \Psi(t,x)}{\partial t}\chi(t,x) - \Psi(t,x)\frac{\partial \chi(t,x)}{\partial t},
$$

$$
\Delta_L \Phi(t,x) = \Delta_L V\left(\chi(t,x) - \frac{\|x\|_H^2}{2}, x\right) - \Delta_L \Psi(t,x)\chi(t,x)
$$
$$
- \Psi(t,x)\Delta_L \chi(t,x).
$$

In so far as

$$
\frac{\partial V\left(\chi(t,x) - \frac{\|x\|_H^2}{2}, x\right)}{\partial t} = \frac{\partial V(\tau,x)}{\partial \tau}\bigg|_{\tau=\chi(t,x)-\frac{\|x\|_H^2}{2}}\frac{\partial \chi(t,x)}{\partial t}
$$
$$
= \Psi(t,x)\frac{\partial \chi(t,x)}{\partial t},
$$

we have

$$
\frac{\partial \Phi(t,x)}{\partial t} = -\chi(t,x)\frac{\partial \Psi(t,x)}{\partial t}. \tag{7.8}
$$

In so far as

$$
\Delta_L V\left(\chi(t,x) - \frac{\|x\|_H^2}{2}, x\right)
$$
$$
= \frac{\partial V(\tau,x)}{\partial \tau}\bigg|_{\tau=\chi(t,x)-\frac{\|x\|_H^2}{2}}[\Delta_L \chi(t,x) - 1]
$$
$$
+ \Delta_L V(\tau,x)\big|_{\tau=\chi(t,x)-\frac{\|x\|_H^2}{2}} = \Psi(t,x)\Delta_L \chi(t,x)
$$
$$
- \left[\frac{\partial V(\tau,x)}{\partial t} - \Delta_L V(\tau,x)\right]\bigg|_{\tau=\chi(t,x)-\frac{\|x\|_H^2}{2}}
$$
$$
= \Psi(t,x)\Delta_L \chi(\tau,x),
$$

since, by condition (7.4) of the theorem $(\partial V(\tau, x)/\partial t) = \Delta_L V(\tau, x)$, we have

$$\Delta_L \Phi(t, x) = -\chi(t, x)\Delta_L \Psi(t, x). \tag{7.9}$$

Substituting (7.8) into (7.6), and (7.9) into (7.7) and taking into account the condition (7.3) of the theorem we derive

$$\frac{\partial U(t, x)}{\partial t} = f(t, \Psi(t, x))$$

$$+ \left[f_\zeta'(\Psi(t, x))t + \frac{\|x\|_H^2}{2} - \chi(t, x) \right] \frac{\partial \Psi(t, x)}{\partial t} = f(\Psi(t, x)),$$

$$\Delta_L U(t, x) = \Psi(t, x))$$

$$+ \left[f_\zeta'(\Psi(t, x))t + \frac{\|x\|_H^2}{2} - \chi(t, x) \right] \Delta_L \Psi(t, x) = \Psi(t, x).$$

Substituting these expressions into equation (7.1), we obtain the identity. Putting $t = 0$ in (7.5) and taking into account that $\chi(0, x) = \|x\|_H^2/2$ and hence, $\Phi(0, x) = V(\chi(0, x) - \|x\|_H^2/2, x) - \Psi(0, x)$ and $\|x\|_H^2/2 = V(0, x) - \Psi(0, x)\|x\|_H^2/2$, we obtain

$$U(0, x) = \Psi(0, x)\frac{\|x\|_H^2}{2} + \Phi(0, x) = V(0, x) = U_0(x).$$

The final statement of the theorem is obvious. □

Example 7.1 Let us solve the Cauchy problem

$$\frac{\partial U(t, x)}{\partial t} = \ln \Delta_L U(t, x) \quad (\Delta_L U(t, x) > 0), \quad U(0, x) = (Ax, x)_H^2,$$

where $A = E + S$, E is the identity operator, and S is some self-adjoint compact operator in H. For this equation,

$$f(\zeta) = \ln \zeta,$$

and the solution of the Cauchy problem

$$\frac{\partial V(\tau, x)}{\partial \tau, x} = \Delta_L V(\tau, x), \quad V(0, x) = (Ax, x)_H^2$$

has the form (see Section 5.4)

$$V(\tau, x) = [2\tau + (Ax, x)_H]^2.$$

Therefore, (7.3) takes the form

$$X^2 - \frac{1}{2}\left[\frac{\|x\|_H^2}{2} - (Sx, x)_H\right]X - \frac{1}{4}\frac{\|x\|_H^2}{2}(Sx, x)_H - \frac{t}{8} = 0,$$

and its solution is given by

$$\chi(t, x) = \frac{1}{4}\left[\frac{||x||_H^2}{2} - (Sx, x)_H + \sqrt{(Ax, x)_H^2 + 2t}\right].$$

From (7.5) we derive the solution of the problem:

$$U(t, x) = t \ln 2\left[(Ax, x)_H + \sqrt{(Ax, x)_H^2 + 2t}\right]$$
$$- \frac{1}{4}\left[(Ax, x)_H - \sqrt{(Ax, x)_H^2 + 2t}\right]^2 + (Ax, x)_H^2.$$

Let us consider a nonlinear equation

$$\frac{\partial U(t, x)}{\partial t} = f(t, \Delta_L U(t, x)), \tag{7.10}$$

where $U(t, x)$ is a function on $[0, T] \times H$, and $f(t, \zeta)$ is a given function of two variables.

Theorem 7.2 *Let $f(t, \zeta)$ be some function of two variables twice continuously differentiable with respect to ζ and continuous in the domain $[0, T] \times \{\Delta_L U(t, x)\}$ in $\mathbb{R}^2$; $\{\Delta_L U(t, x)\}$ is a range in $\mathbb{R}^1$.*

1. Then the solution of equation (7.10) has the form

$$U(t, x) = \int_0^t f(s, \Psi(x))\, ds + \Psi(x)\frac{||x||_H^2}{2} + \Phi(x), \tag{7.11}$$

where $\Psi(x)$, $\Phi(x)$ are arbitrary harmonic functions on H.
2. Let the equation

$$\int_0^t f_\zeta'\left(s, \left.\frac{\partial V(\tau, x)}{\partial \tau}\right|_{\tau = X - \frac{||x||_H^2}{2}}\right) ds + \frac{||x||_H^2}{2} - X = 0, \tag{7.12}$$

where $V(\tau, x)$ is the solution of the Cauchy problem for the 'heat equation'

$$\frac{\partial V(\tau, x)}{\partial \tau} = \Delta_L V(\tau, x), \quad V(0, x) = U_0(x) \tag{7.13}$$

for some given function $U_0(x)$ (and $\partial^2 V(\tau, x)/\partial \tau^2 \neq 0$), be solvable with respect to $X = \chi(t, x)$ (and $\chi(0, x) = (||x||_H^2/2)$). Then the solution of the Cauchy problem

$$\frac{\partial U(t, x)}{\partial t} = f(t, \Delta_L U(x)), \quad U(0, x) = U_0(x)$$

for the nonlinear equation (7.10) can be determined according to the formula

$$U(t, x) = \int_0^t f(s, \Psi(t, x)) \, ds + \Psi(t, x) \frac{||x||_H^2}{2} + \Phi(t, x), \qquad (7.14)$$

where

$$\Psi(t, x) = \frac{\partial V(\tau, x)}{\partial \tau} \Big|_{\tau = \chi(t,x) - \frac{||x||_H^2}{2}},$$

$$\Phi(t, x) = V(\chi(t, x) - \frac{||x||_H^2}{2}, x) - \Psi(t, x)\chi(t, x);$$

$V(\tau, x)$ *is the solution of the Cauchy problem* (7.13).

If, in addition, equation (7.12) *is uniquely solvable with respect to* $\chi(t, x)$ *and if the Cauchy problem for the 'heat equation' has a unique solution in some functional class, then the solution of the Cauchy problem for equation* (7.10) *is unique in the same class.*

Proof. 1. From (7.11), by (1.4) and by harmonicity of $\Psi(x)$ and $\Phi(x)$, we obtain

$$\frac{\partial U(t, x)}{\partial t} = f(t, \Psi(x)),$$

$$\Delta_L U(t, x) = \int_0^t f'_\zeta(s, \Psi(x)) \, ds \, \Delta_L \Psi(x) + \frac{||x||_H^2}{2} \Delta_L \Psi(x)$$

$$+ \Psi(x) + \Delta_L \Phi(x) = \Psi(x).$$

Substituting these expressions into (7.10), we obtain the identity.
2. From (7.14), by (1.4), we obtain

$$\frac{\partial U(t, x)}{\partial t} = f(t, \Psi(t, x)) + \int_0^t f'_\zeta(s, \Psi(t, x)) \, ds \frac{\partial \Psi(t, x)}{\partial t}$$

$$+ \frac{||x||_H^2}{2} \frac{\partial \Psi(t, x)}{\partial t} + \frac{\partial \Phi(t, x)}{\partial t}, \qquad (7.15)$$

$$\Delta_L U(t, x) = \int_0^t f'_\zeta(\Psi(t, x)) \, ds \, \Delta_L \Psi(t, x)$$

$$+ \frac{||x||_H^2}{2} \Delta_L \Psi(t, x) + \Psi(t, x) + \Delta_L \Phi(t, x). \qquad (7.16)$$

Substituting the expressions for $\partial \Phi(t, x)/\partial t$ and $\Delta_L \Phi(t, x)$ into (7.15), (7.16) (formulae (7.8), (7.9)) and taking into account condition (7.12) of the theorem, we obtain

$$\frac{\partial U(t, x)}{\partial t} = f(t, \Psi(t, x))$$

$$+ \left[\int_0^t f'_\zeta(s, \Psi(t, x)) \, ds + \frac{||x||_H^2}{2} - \chi(t, x) \right] \frac{\partial \Psi(t, x)}{\partial t}$$

$$= f(t, \Psi(t, x)),$$

$$\Delta_L U(t, x)$$

$$= \Psi(t, x) + \left[\int_0^t f'_\zeta(s, \Psi(t, x)) \, ds + \frac{||x||_H^2}{2} - \chi(t, x) \right] \Delta_L \Psi(t, x)$$

$$= \Psi(t, x).$$

Substituting these expressions into equation (7.10), we obtain an identity.

Putting $t = 0$ in (7.14) and taking into account that $\chi(0, x) = ||x||_H^2/2$, we obtain

$$U(0, x) = \Psi(0, x) \frac{||x||_H^2}{2} + \Phi(0, x) = V\left(\chi(0, x) - \frac{||x||_H^2}{2}, x\right)$$

$$= V(0, x) = U_0(x).$$

The final statement of the theorem is obvious.   □

Example 7.2 Let us solve the Cauchy problem

$$\frac{\partial U(t, x)}{\partial t} = 4(t + 1)e^{-\frac{1}{2}\Delta_L U(t, x)},$$

$$U(0, x) = ||x||_H^2 \left(1 - \ln \frac{||x||_H^2}{2}\right).$$

Here

$$f(t, \zeta) = 4(t + 1)e^{-\frac{1}{2}\zeta},$$

and the solution of the Cauchy problem

$$\frac{\partial V(\tau, x)}{\partial t} = \Delta_L V(\tau, x), \quad V(0, x) = ||x||_H^2 \left(1 - \ln \frac{||x||_H^2}{2}\right)$$

has the form (see Section 5.4)

$$V(\tau, x) = (2\tau + ||x||_H^2)\left[1 - \ln\left(\tau + \frac{||x||_H^2}{2}\right)\right].$$

Therefore, (7.12) takes the form

$$(1 + t)^2 X - \frac{||x||_H^2}{2} = 0,$$

and its solution is given by

$$\chi(t, x) = \frac{||x||_H^2}{2(1+t)^2}.$$

From (7.14) we derive the solution of the problem:

$$U(t, x) = ||x||_H^2 \left[1 - \ln \frac{||x||_H^2}{2(1+t)^2} \right].$$

7.2 The Cauchy problem for the equation $(\partial U(t,x)/\partial t) = f(U(t,x), \Delta_L U(t,x))$

Consider the nonlinear equation

$$\frac{\partial U(t, x)}{\partial t} = f(U(t, x), \Delta_L U(t, x)), \tag{7.17}$$

where $U(t, x)$ is a function on $[0, T] \times H$, and $f(\xi, \zeta)$ is a given function of two variables.

Theorem 7.3 *Let* $f(\xi, \zeta)$ *be a twice continuously differentiable function of two variables in the range of* $\{U(t, x), \Delta_L U(t, x)\}$ *in* $\mathbb{R}^2$. *Let the equation* $\eta = f(\xi, c\eta)$ *be solvable with respect to* η, $\eta = \phi(\xi, c)$, *and the variables* ξ *and* c *be separable:* $\phi(\xi, c) = \alpha(c)\beta(\xi)$ *(here* $\alpha(c)$, $\beta(\xi)$ *are functions on* $\mathbb{R}^1$, $\beta(\xi) \neq 0$).

1. *Then the solution of equation (7.17) (written implicitly) has the form*

$$\varphi(U(t, x)) = \alpha(\Psi(x)) \left[t + \Psi(x) \frac{||x||_H^2}{2} \right] + \Phi(x), \tag{7.18}$$

where $\varphi(\xi) = \int (d\xi / \beta(\xi))$, $\Psi(x)$, $\Phi(x)$ *are arbitrary harmonic functions on* H.

2. *Let the equation*

$$\alpha_c' \left(\frac{\partial V(\tau, x)/\partial \tau}{f(V(\tau, x), \partial V(\tau, x)/\partial \tau)} \Big|_{\tau = X - \frac{||x||_H^2}{2}} \right) t$$

$$+ \gamma_c' \left(\frac{\partial V(\tau, x)/\partial \tau}{f(V(\tau, x), \partial V(\tau, x)/\partial \tau)} \Big|_{\tau = X - \frac{||x||_H^2}{2}} \right) \left[\frac{||x||_H^2}{2} - X \right] = 0, \tag{7.19}$$

where $\gamma(c) = c\,\alpha(c)$, $V(\tau, x)$, *is the solution of the Cauchy problem for the 'heat equation'*

$$\frac{\partial V(\tau, x)}{\partial \tau} = \Delta_L V(\tau, x), \quad V(0, x) = U_0(x), \tag{7.20}$$

$(\partial^2 V(\tau, x)/\partial\tau^2 \neq 0)$, *is solvable with respect to* $X = \chi(t, x)$ *(and* $\chi(0, x) = \|x\|_H^2/2$*). Let the function* φ *exist and have an inverse* φ^{-1}. *Then the solution of the Cauchy problem*

$$\frac{\partial U(t, x)}{\partial t} = f(U(t, x), \Delta_L U(x)), \quad U(0, x) = U_0(x)$$

for the nonlinear equation (7.17) can be determined by

$$\varphi(U(t, x)) = \alpha(\Psi(t, x))\left[t + \Psi(t, x)\frac{\|x\|_H^2}{2}\right] + \Phi(t, x), \qquad (7.21)$$

where

$$\Psi(t, x) = \left.\frac{\dfrac{\partial V(\tau, x)}{\partial \tau}}{f\left(V(t, x), \dfrac{\partial V(\tau, x)}{\partial \tau}\right)}\right|_{\tau = \chi(t, x) - \frac{1}{2}\|x\|_H^2},$$

$$\Phi(t, x) = \varphi\left(V\left(\chi(t, x) - \frac{\|x\|_H^2}{2}, x\right)\right) - \gamma(\Psi(t, x))\chi(t, x);$$

$V(t, x)$ *is solution of the Cauchy problem (7.20).*

If, in addition, the equations $\eta = f(\xi, c\eta)$ *and (7.19) are uniquely solvable, respectively, with respect to* η *and* $\chi(t, x)$ *and if the Cauchy problem for the 'heat equation' has a unique solution in some functional class, then the solution of the Cauchy problem for equation (7.17) is unique in the same class.*

Proof. 1. From (7.18), by (1.4), we derive

$$\varphi'_\xi(U(t, x))\frac{\partial U(t, x)}{\partial t} = \alpha(\Psi(x)),$$

$$\varphi'_\xi(U(t, x))\Delta_L U(t, x) = \alpha'_c(\Psi(x))\Delta_L\Psi(x)\left[t + \Psi(x)\frac{\|x\|_H^2}{2}\right]$$

$$+ \alpha(\Psi(x))\left[\frac{\|x\|_H^2}{2}\Delta_L\Psi(x) + \frac{1}{2}\Psi(x)\Delta_L\|x\|_H^2\right]$$

$$+ \Delta_L\Phi(x) = \Psi(x)\alpha(\Psi(x))$$

(since $\Delta_L\Psi(x) = \Delta_L\Phi(x) = 0$, $\Delta_L\|x\|_H^2 = 2$). Hence

$$\frac{\partial U(t, x)}{\partial t} = \alpha(\Psi(x))\beta(U(t, x)),$$

$$\Delta_L U(t, x) = \Psi(x)\alpha(\Psi(x))\beta(U(t, x))$$

(since $\varphi'_\xi(\xi) = 1/\beta(\xi)$).

Substituting these expressions into (7.17), we obtain the identity

$$\alpha(\Psi(x))\beta(U(t, x)) = f(U(t, x), \Psi(x)\alpha(\Psi(x))\beta(U(t, x))),$$

since by the conditions of the theorem the equality $\eta = \alpha(c)\beta(\xi)$ follows from $\eta = f(\xi, c\eta)$.

2. From (7.21), by (1.4), we derive

$$\varphi'_\xi(U(t,x))\frac{\partial U(t,x)}{\partial t} = \alpha(\Psi(t,x)) + \alpha(\Psi(t,x))\frac{\|x\|^2_H}{2}\frac{\partial \Psi(t,x)}{\partial t}$$
$$+ \alpha'_c(\Psi(t,x))\frac{\partial \Psi(t,x)}{\partial t}\left[t + \Psi(t,x)\frac{\|x\|^2_H}{2}\right] + \frac{\partial \Phi(t,x)}{\partial t}, \qquad (7.22)$$

$$\varphi'_\xi(U(t,x))\Delta_L U(t,x) = \alpha'_c(\Psi(t,x))\Delta_L\Psi(t,x)\left[t + \Psi(t,x)\frac{\|x\|^2_H}{2}\right]$$
$$+ \alpha(\Psi(t,x))\frac{\|x\|^2_H}{2}\Delta_L\Psi(t,x) + \Psi(t,x)\alpha(\Psi(t,x)) + \Delta_L\Phi(t,x). \qquad (7.23)$$

By direct computation we derive expressions for $\partial\Phi(t,x)/\partial t$ and $\Delta_L\Phi(t,x)$:

$$\frac{\partial\Phi(t,x)}{\partial t} = \frac{1}{\beta\left(V\left(\chi(t,x) - \|x\|^2_H/2, x\right)\right)}\frac{\partial V\left(\chi(t,x) - \|x\|^2_H/2, x\right)}{\partial t}$$
$$-\gamma'_c(\Psi(t,x))\frac{\partial\Psi(t,x)}{\partial t}\chi(t,x) - \gamma(\Psi(t,x))\frac{\partial\chi(t,x)}{\partial t},$$

$$\Delta_L\Phi(t,x) = \frac{1}{\beta\left(V\left(\chi(t,x) - \frac{\|x\|^2_H}{2}, x\right)\right)}\Delta_L V\left(\chi(t,x) - \frac{\|x\|^2_H}{2}, x\right)$$
$$-\gamma'_c(\Psi(t,x))\Delta_L\Psi(t,x)\chi(t,x) - \gamma(\Psi(t,x))\Delta_L\chi(t,x).$$

In addition,

$$\frac{\partial V\left(\chi(t,x) - \frac{\|x\|^2_H}{2}, x\right)}{\partial t} = \left.\frac{\partial V(\tau,x)}{\partial \tau}\right|_{\tau=\chi(t,x)-\frac{1}{2}\|x\|^2_H}\frac{\partial\chi(t,x)}{\partial t}.$$

It follows from the condition of the theorem that

$$\Psi(t,x) = \left.\frac{\frac{\partial V(\tau,x)}{\partial\tau}}{f\left(V(t,x), \frac{\partial V(\tau,x)}{\partial\tau}\right)}\right|_{\tau=\chi(t,x)-\frac{1}{2}\|x\|^2_H},$$

and hence

$$\left.\frac{\partial V(\tau,x)}{\partial\tau}\right|_{\tau=\chi(t,x)-\frac{1}{2}\|x\|^2_H} = \beta\left(V\left(\chi(t,x) - \frac{\|x\|^2_H}{2}, x\right)\right)\Psi(t,x)\alpha(\Psi(t,x)),$$

(since from $\eta = f(\xi, c\eta)$ it follows that $\eta = \alpha(c)\beta(\xi)$). Hence,

$$\frac{\partial V\left(\chi(t, x) - \frac{\|x\|_H^2}{2}, x\right)}{\partial t} = \gamma(\Psi(t, x))\beta\left(V\left(\chi(t, x) - \frac{\|x\|_H^2}{2}, x\right)\right)\frac{\partial \chi(t, x)}{\partial t}.$$

Therefore

$$\frac{\partial \Phi(t, x)}{\partial t} = -\gamma_c'(\Psi(t, x))\chi(t, x)\frac{\partial \Psi(t, x)}{\partial t}. \tag{7.24}$$

Furthermore,

$$\Delta_L V(\chi(t, x) - \frac{1}{2}\|x\|_H^2, x)$$

$$= \frac{\partial V(\tau, x)}{\partial \tau}\bigg|_{\tau=\chi(t,x)-\frac{1}{2}\|x\|_H^2} [\Delta_L \chi(t, x) - 1] + \Delta_L V(\tau, x)\bigg|_{\tau=\chi(t,x)-\frac{1}{2}\|x\|_H^2}$$

$$= \frac{\partial V(\tau, x)}{\partial \tau}\bigg|_{\tau=\chi(t,x)-\frac{1}{2}\|x\|_H^2} \Delta_L \chi(t, x)$$

$$- \left[\frac{\partial V(\tau, x)}{\partial \tau} - \Delta_L V(\tau, x)\right]\bigg|_{\tau=\chi(t,x)-\frac{1}{2}\|x\|_H^2}$$

$$= \frac{\partial V(\tau, x)}{\partial \tau}\bigg|_{\tau=\chi(t,x)-\frac{1}{2}\|x\|_H^2}$$

$$\times \Delta_L \chi(t, x) = \beta\left(V\left(\chi(t, x) - \frac{\|x\|_H^2}{2}, x\right)\right)\Psi(t, x)\alpha(\Psi(t, x))\Delta_L \chi(t, x),$$

which yields

$$\Delta_L \Phi(t, x) = -\gamma_c'(\Psi(t, x))\chi(t, x)\Delta_L \Psi(t, x). \tag{7.25}$$

Substituting (7.24) into (7.22), and (7.25) into (7.23), we obtain, by condition (7.19) of the theorem,

$$\frac{1}{\beta(U(t, x))}\frac{\partial U(t, x)}{\partial t} = \alpha(\Psi(t, x)) + \left\{\alpha_c'(\Psi(t, x))t\right.$$

$$\left. + \gamma_c'(\Psi(t, x))\left[\frac{\|x\|_H^2}{2} - \chi(t, x)\right]\right\}\frac{\partial \Psi(t, x)}{\partial t}$$

$$= \alpha(\Psi(t, x)),$$

$$\frac{1}{\beta(U(t, x))}\Delta_L U(t, x) = \gamma(\Psi(t, x))$$

$$+ \left\{\alpha_c'(\Psi(t, x))t + \gamma_c'(\Psi(t, x))\left[\frac{\|x\|_H^2}{2} - \chi(t, x)\right]\right\}$$

$$\times \Delta_L(\Psi(t, x)) = \gamma(t, x).$$

Hence

$$\frac{\partial U(t, x)}{\partial t} = \alpha(\Psi(t, x))\beta(U(t, x)), \quad \Delta_L U(t, x)) = \gamma(\Psi(t, x))\beta(U(t, x)).$$

Substituting these expressions into equation (7.17) and taking into account that by the conditions of the theorem the equality $\eta = f(\xi, c\eta)$ yields $\eta = \alpha(c)\beta(\xi)$, we obtain the identity

$$\alpha(\Psi(t,x))\beta(U(t,x)) = f(U(t,x), \gamma(\Psi(t,x))\beta(U(t,x))).$$

Putting $t = 0$ in (7.21) and taking into account that $\chi(0,x) = 1/2\|x\|_H^2$, and, therefore, $\Phi(0,x) = \varphi(V(\chi(0,x) - \|x\|_H^2/2, x)) - \gamma(\Psi(0,x))\|x\|_H^2/2 = \varphi(V(0,x)) - \gamma(\Psi(0,x))\|x\|_H^2/2$, we obtain

$$\varphi(U(0,x)) = \Psi(0,x)\alpha(\Psi(0,x))\frac{\|x\|_H^2}{2} + \Phi(0,x)$$
$$= \varphi(V(0,x)) = \varphi(U_0(x))$$

and $U(0,x) = U_0(x)$.

The final statement of the theorem is obvious. $\qquad\qquad\square$

Example 7.3 Let us solve the Cauchy problem

$$\frac{\partial U(t,x)}{\partial t} = \frac{[\Delta_L U(t,x)]^2}{U(t,x)} \quad (U(t,x) \neq 0), \quad U(0,x) = \|x\|_H^4.$$

Here

$$f(\xi, \zeta) = \zeta^2/\xi, \quad \alpha(c) = 1/c^2, \quad \beta(\xi) = \xi, \quad \varphi(\xi) = \ln|\xi|,$$

and the solution of the Cauchy problem

$$\frac{\partial V(\tau,x)}{\partial \tau} = \Delta_L V(\tau,x), \quad V(0,x) = \|x\|_H^4$$

has the form (see Section 5.4)

$$V(\tau,x) = 4\left[\tau + \frac{\|x\|_H^2}{2}\right]^2.$$

Therefore, (7.19) takes the form

$$X^2 - \frac{1}{2}\|x\|_H^2 X - 4t = 0,$$

and its solution is given by

$$\chi(t,x) = \frac{1}{4}\|x\|_H^2 + \frac{1}{2}\sqrt{\frac{1}{4}\|x\|_H^4 + 16t}.$$

From (7.21), we derive the expression for the solution of the problem

$$U(t,x) = \left[\frac{\|x\|_H^2}{2} + \sqrt{\frac{\|x\|_H^4}{4} + 16t}\right]^2$$

$$\times \exp\left\{-\frac{16t}{\left[\frac{\|x\|_H^2}{2} + \sqrt{\frac{\|x\|_H^4}{4} + 16t}\right]^2}\right\}.$$

The solution of the Cauchy problem for the quasilinear equation

$$\frac{\partial U(t, x)}{\partial t} = \Delta_L U(t, x) + f_0(U(t, x)), \tag{7.26}$$

where $f_0(\xi)$ is a given function of one variable has an especially simple form.

By Theorem 7.3, we have

Corollary. *The solution of the Cauchy problem*

$$\frac{\partial U(t, x)}{\partial t} = \Delta_L U(t, x) + f_0(U(t, x)), \quad U(0, x) = U_0(x)$$

for the quasilinear equation (7.26) *has the form*

$$\varphi(U(t, x)) = t + \varphi(V(t, x)), \tag{7.27}$$

where $\varphi(\xi) = \int (d\xi / f_0(\xi))$, *and* $V(t, x)$ *is the solution of the Cauchy problem for the 'heat equation'*

$$\frac{\partial V(t, x)}{\partial t} = \Delta_L V(t, x), \quad V(0, x) = U_0(x).$$

Indeed, since

$$f(\xi, \zeta) = \zeta + f_0(\xi),$$

in (7.26), it follows that $\alpha(c) = 1/(1 - c)$, $\beta(\xi) = f_0(\xi)$.

In so far as we have $\alpha'_c(c) = \gamma'_c = 1/(1 - c)^2$, (7.19) takes the form

$$t + \frac{1}{2}||x||_H^2 - X = 0,$$

and hence

$$\chi(t, x) = t + \frac{1}{2}||x||_H^2.$$

By (7.21) we have

$$\varphi(U(t, x)) = \alpha(\Psi(t, x))t + \gamma(\Psi(t, x)) \left[\frac{1}{2}||x||_H^2 - \chi(t, x) \right]$$

$$+ \varphi \left(V \left(\chi(t, x) - \frac{1}{2}||x||_H^2, x \right) \right).$$

In so far as

$$\alpha(\Psi(t, x)) = \frac{1}{1 - \Psi(t, x)}, \quad \gamma(\Psi(t, x)) = \frac{\Psi(t, x)}{1 - \Psi(t, x)},$$

$$\chi(t, x) = t + \frac{1}{2}||x||_H^2, \quad V \left(\chi(t, x) - \frac{1}{2}||x||_H^2, x \right) = V(t, x),$$

we have

$$\varphi(U(t, x)) = t + \varphi(V(t, x)).$$

$\square$

Example 7.4 Let us solve the Cauchy problem

$$\frac{\partial U(t, x)}{\partial t} = \Delta_L U(t, x) - U^m(x), \quad m > 1,$$

$$U(x, 0) = (Ax, x)_H^2,$$

where $A = E + S$, E is the identity operator, and S is a self-adjoint compact operator in H.

Here

$$f_0(\xi) = -\xi^m,$$

therefore

$$\varphi(\xi) = \frac{1}{(m-1)\xi^{m-1}}.$$

The solution of the Cauchy problem

$$\frac{\partial V(\tau, x)}{\partial \tau} = \Delta_L V(\tau, x), \quad V(x, 0) = (Ax, x)_H^2$$

has the form (see Section 5.4)

$$V(\tau, x) = [2\tau + (Ax, x)_H]^2.$$

By (7.27) we derive the solution of the Cauchy problem:

$$U(t, x) = \frac{[2t + (Ax, x)_H]^2}{\{(m-1)t[2t + (Ax, x)_H]^{2m-2} + 1\}^{\frac{1}{m-1}}}.$$

7.3 The Cauchy problem for the equation $\varphi(t, \partial U(t, x)/\partial t) = f(F(x), \Delta_L U(t, x))$

Consider the nonlinear equation

$$\varphi\left(t, \frac{\partial U(t, x)}{\partial t}\right) = f(F(x), \Delta_L U(t, x)), \quad (7.28)$$

where $U(t, x)$ defined on $[0, T] \times H$; $\varphi(t, \xi)$, and $f(\alpha, \zeta)$ are given functions of two variables, and $F(x)$ is a given function on H.

Theorem 7.4 *Let $\varphi(t, \xi)$ and $f(\alpha, \zeta)$ be functions of two variables continuous, respectively, in the domain $[0, T] \times \{\partial U(t, x)/\partial t\}$ (here $\{\partial U(t, x)/\partial t\}$ is the range of $\partial U(t, x)/\partial t$ in $\mathbb{R}^1$) and in the range of $\{F(x), \Delta_L U(t, x)\}$ in $\mathbb{R}^2$. Let in addition, the equations $\varphi(t, \xi) = a$, $f(c\alpha, \zeta) = a$ be solvable, respectively,* *with respect to ξ and ζ, i.e., $\xi = h(t, a)$ and $\zeta = g(\alpha, a)$. Assume that $h(t, a)$ and $g(\alpha, a)$ and the partial derivatives $h'_a(t, a)$ and $g'_a(\alpha, a)$ are continuous in their domains of definition, and $F(x)$ is such that $\Delta_L F(x) = c \neq 0$, where c is a constant.*

1. *Then the solution of equation (7.28) has the form*

$$U(t, x) = \int_0^t h(s, \Psi(x))ds + \int_0^{\frac{1}{c}F(x)} g(\alpha, \Psi(x))d\alpha + \Phi(x), \qquad (7.29)$$

where $\Psi(x)$, $\Phi(x)$ are arbitrary harmonic functions on H.
2. *Let the equation*

$$\int_0^t h'_a\left(s, f\left(X, \left.\frac{\partial V(\tau, x)}{\partial \tau}\right|_{\tau = X - \frac{1}{c}F(x)}\right)\right)ds$$

$$+ \int_X^{\frac{1}{c}F(x)} g'_a\left(\alpha, f\left(X, \left.\frac{\partial V(\tau, x)}{\partial \tau}\right|_{\tau = X - \frac{1}{c}F(x)}\right)\right)d\alpha = 0, \qquad (7.30)$$

where $V(\tau, x)$ is the solution of the Cauchy problem for the 'heat equation'

$$\frac{\partial V(\tau, x)}{\partial \tau} = \Delta_L V(\tau, x), \quad V(0, x) = U_0(x) \qquad (7.31)$$

be solvable with respect to $X = \chi(t, x)$, with $\chi(0, x) = \frac{1}{c}F(x)$. Then the solution of the Cauchy problem

$$\varphi\left(t, \frac{\partial U(t, x)}{\partial t}\right) = f(F(x), \Delta_L U(t, x)), \; U(0, x) = U_0(x)$$

for the nonlinear equation (7.28) has the form

$$U(t, x) = \int_0^t h(s, \Psi(t, x))ds + \int_{\chi(t,x)}^{\frac{1}{c}F(x)} g(\alpha, \Psi(t, x))d\alpha + \Phi(t, x), \quad (7.32)$$

where

$$\Psi(t, x) = f\left(\chi(t, x), \left.\frac{\partial V(\tau, x)}{\partial \tau}\right|_{\tau = \chi(t,x) - \frac{1}{c}F(x)}\right),$$

$$\Phi(t, x) = V\left(\chi(t, x) - \frac{1}{c}F(x), x\right);$$

$V(t, x)$ is the solution of Cauchy problem (7.31).

If, in addition, the equations $\varphi(t, \xi) = \alpha$, $f(c\alpha, \zeta) = \alpha$ and (7.30) are uniquely solvable, respectively, with respect to ξ, ζ and $\chi(t, x)$ and if the Cauchy problem for the 'heat equation' has a unique solution in some functional class, then the solution of the Cauchy problem for equation (7.28) is unique in the same class.

Proof. 1. From (7.29) by (1.4), we obtain

$$\frac{\partial U(t, x)}{\partial t} = h(t, \Psi(x)),$$

$$\Delta_L U(t, x) = \int_0^t h'_a(s, \Psi(x))\, ds\, \Delta_L \Psi(x)$$

$$+ \int_0^{\frac{1}{c}F(x)} g'_a(\alpha, \Psi(x))\, d\alpha\, \Delta_L \Psi(x)$$

$$+ g\left(\frac{1}{c}F(x), \Psi(x)\right)\frac{1}{c}\Delta_L F(x) + \Delta_L \Phi(x).$$

But by harmonicity $\Delta_L \Psi(x) = \Delta_L \Phi(x) = 0$, and by the conditions of the theorem, $\Delta_L F(x) = c$; therefore

$$\Delta_L U(t, x) = g\left(\frac{1}{c}F(x), \Psi(x)\right).$$

Substituting the expressions $\partial U(t, x)/\partial t$ and $\Delta_L U(t, x)$ into (7.28), we obtain the identity

$$\varphi(t, h(t, \Psi(x))) = f\left(F(x), g\left(\frac{1}{c}F(x), \Psi(x)\right)\right),$$

since by the condition of the theorem, $\varphi(t, \xi) = f(c\alpha, \zeta)$.

2. From (7.32) by (1.4), we obtain

$$\frac{\partial U(t, x)}{\partial t} = h(t, \Psi(t, x)) + \int_0^t h'_a(s, \Psi(t, x))\, ds\, \frac{\partial \Psi(t, x)}{\partial t}$$

$$+ \int_{\chi(t,x)}^{\frac{1}{c}F(x)} g'_a(\alpha, \Psi(t, x))\, d\alpha\, \frac{\partial \Psi(t, x)}{\partial t}$$

$$- g(\chi(t, x), \Psi(t, x))\frac{\partial \chi(t, x)}{\partial t} + \frac{\partial \Phi(t, x)}{\partial t}, \tag{7.33}$$

$$\Delta_L U(t, x) = \int_0^t h'_a(s, \Psi(t, x))\, ds\, \Delta_L \Psi(t, x)$$

$$+ \int_{\chi(t,x)}^{\frac{1}{c}F(x)} g'_a(\alpha, \Psi(t, x))\, d\alpha\, \Delta_L \Psi(t, x)$$

$$+ g\left(\frac{1}{c}F(x), \Psi(t, x)\right)\frac{1}{c}\Delta_L F(x)$$

$$- g(\chi(t, x), \Psi(t, x))\Delta_L \chi(t, x) + \Delta_L \Phi(t, x). \tag{7.34}$$

Let us compute $\partial \Phi(t, x)/\partial t$:

$$\frac{\partial \Phi(t, x)}{\partial t} = \frac{\partial V(\tau, x)}{\partial \tau}\bigg|_{\tau = \chi(t,x) - \frac{1}{c}F(x)} \frac{\partial \chi(t, x)}{\partial t}.$$

But the relation $g(\alpha, f(\alpha, \zeta)) = \zeta$ in the condition of the theorem yields

$$\frac{\partial V(\tau, x)}{\partial \tau}\bigg|_{\tau = \chi(t,x) - \frac{1}{c}F(x)} = g\left(\chi(t, x), f\left(\chi(t, x), \frac{\partial V(\tau, x)}{\partial \tau}\bigg|_{\tau = \chi(t,x) - \frac{1}{c}F(x)}\right)\right)$$
$$= g(\chi(t, x), \Psi(t, x)).$$

Therefore

$$\frac{\partial \Phi(t, x)}{\partial t} = g(\chi(t, x), \Psi(t, x)) \frac{\partial \chi(t, x)}{\partial t}. \tag{7.35}$$

Let us compute $\Delta_L \Phi(t, x)$

$$\Delta_L \Phi(t, x) = \frac{\partial V(\tau, x)}{\partial \tau}\bigg|_{\tau = \chi(t,x) - \frac{1}{c}F(x)} \left[\Delta_L \chi(t, x) - \frac{1}{c}\Delta_L F(x)\right]$$
$$+ \Delta_L V(\tau, x)\bigg|_{\tau = \chi(t,x) - \frac{1}{c}F(x)}.$$

But from $g(\alpha, f(\alpha, \zeta)) = \zeta$ it follows that

$$\frac{\partial V(\tau, x)}{\partial \tau}\bigg|_{\tau = \chi(t,x) - \frac{1}{c}F(x)} = g\left(\chi(t, x), f\left(\chi(t, x), \frac{\partial V(\tau, x)}{\partial \tau}\bigg|_{\tau = \chi(t,x) - \frac{1}{c}F(x)}\right)\right)$$
$$= g(\chi(t, x), \Psi(t, x)).$$

Therefore

$$\Delta_L \Phi(t, x) = g(\chi(t, x), \Psi(t, x)) \Delta_L \chi(t, x)$$
$$- \left[\frac{\partial V(\tau, x)}{\partial \tau} - \Delta_L V(\tau, x)\right]\bigg|_{\tau = \chi(t,x) - \frac{1}{c}F(x)}.$$

Taking into account that $\partial V(\tau, x)/\partial \tau = \Delta_L V(\tau, x)$, by (7.31) we have

$$\Delta_L \Phi(t, x) = g(\chi(t, x), \Psi(t, x)) \Delta_L \chi(t, x). \tag{7.36}$$

Substituting (7.35) into (7.33), and (7.36) into (7.34), we obtain, taking into account (7.30),

$$\frac{\partial U(t, x)}{\partial t} = h(t, \Psi(t, x)) + \left[\int_0^t h'_a(s, \Psi(t, x)) \, ds\right.$$
$$\left. + \int_{\chi(t,x)}^{\frac{1}{c}F(x)} g'_a(\alpha, \Psi(t, x)) d\alpha\right] \frac{\partial \Psi(t, x)}{\partial t}$$
$$= h(t, \Psi(t, x)),$$

$$\Delta_L U(t, x) = \left[\int_0^t h'_a(s, \Psi(t, x)) \, ds + \int_{\chi(t,x)}^{\frac{1}{c}F(x)} g'_a(\alpha, \Psi(t, x)) d\alpha\right]$$
$$\times \Delta_L \Psi(t, x) + g\left(\frac{1}{c}F(x), \Psi(t, x)\right)$$
$$= g\left(\frac{1}{c}F(x), \Psi(t, x)\right).$$

Substituting these expressions into equation (7.28), we obtain the identity

$$\varphi(t, h(t, \Psi(t, x))) = f\left(\frac{1}{c}F(x), \Psi(t, x)\right).$$

This follows from the condition $\varphi(t, \xi) = f(c\alpha, \zeta)$.

Putting $t = 0$, in (7.32) and taking into account that $\chi(0, x) = \frac{1}{c}F(x)$, we obtain

$$U(0, x) = \Phi(0, x) = V(\chi(0, x) - \frac{1}{c}F(x), x) = V(0, x) = U_0(x).$$

The final statement of the theorem is obvious. $\square$

Example 7.5 Let us solve the Cauchy problem

$$\exp\left\{\frac{\partial U(t, x)}{\partial t}\right\} = (t + 1)[\Delta_L U(t, x) + (Ax, x)_H],$$

$$U(0, x) = \frac{1}{4}(Bx, x)_H^2,$$

where $A = E + S$, $B = E - S$, E is the identity operator, and S is a self-adjoint compact operator in H.

Here

$$F(x) = (Ax, x)_H = ||x||_H^2 + (Sx, x)_H, \quad \Delta_L F(x) = 2,$$

$$\varphi(t, \xi) = \frac{\exp\{\xi\}}{t + 1}, \quad f(c\alpha, \zeta) = 2\alpha + \zeta,$$

and hence

$$h(t, a) = \ln(t + 1)a, \quad g(\alpha, a) = -2\alpha + a.$$

The solution of the Cauchy problem

$$\frac{\partial V(\tau, x)}{\partial \tau} = \Delta_L V(t, x), \quad V(0, x) = \frac{1}{4}(Bx, x)_H^2$$

has the form (see Section 5.4)

$$V(\tau, x) = \left[\tau + \frac{1}{2}(Bx, x)_H\right]^2.$$

Therefore, (7.30) takes the form

$$\left[X - \frac{1}{2}(Sx, x)_H\right]^2 - \frac{1}{2}||x||_H^2\left[X - \frac{1}{2}(Sx, x)_H\right] - \frac{1}{4}t = 0,$$

and its solution is

$$\chi(t, x) = \frac{1}{4}||x||_H^2 + \frac{1}{2}\sqrt{\frac{1}{4}||x||^4 + t} + \frac{1}{2}(Sx, x)_H.$$

By (7.32), we obtain the solution of the problem:

$$U(t, x) = \ln\left[(t + 1)^{t+1}\left(||x||_H^2 + 2\sqrt{\frac{1}{4}||x||_H^4 + t}\right)^t\right] - \frac{3}{2}t$$

$$+ \frac{1}{2}||x||_H^2\sqrt{\frac{1}{4}||x||_H^4 + t} - \frac{1}{2}||x||_H^2(Sx, x)_H + \frac{1}{4}(Sx, x)_H^2.$$

7.4 The Cauchy problem for the equation $f(U(t, x), \partial U(t, x)/\partial t, \Delta_L U(t, x)) = 0$

Consider the nonlinear equation

$$f\left(U(t, x), \frac{\partial U(t, x)}{\partial t}, \Delta_L U(t, x)\right) = 0 \tag{7.37}$$

where $U(t, x)$ is function on $[0, T] \times H$, and $f(\xi, \eta, \zeta)$ is a function of three variables.

Theorem 7.5 *Let the function $f(\xi, \eta, \zeta)$ satisfy the following conditions:*

(1) The function $f(\xi, \eta, \zeta)$ is a twice continuously differentiable function of three variables in the range of $\{U(t, x), \partial U(t, x)/\partial t, \Delta_L U(t, x)\}$ in $\mathbb{R}^3$.
(2) The equation

$$f(\xi, \eta, a\eta) = 0 \tag{7.38}$$

is solvable with respect to η, $\eta = \phi(\xi, a)$ (although the equation $f(\xi, \eta, \zeta) = 0$, generally speaking can not be solved with respect to η), and assume that the variables ξ and a can be separated, i.e. $\phi(\xi, a) = \alpha(a)\beta(\xi)$ where $\alpha(a)$, $\beta(\xi)$ are functions on $\mathbb{R}^1$, $\beta(\xi) \neq 0$.
1. Then the solution of equation (7.37) (written implicitly) has the form

$$\varphi(U(t, x)) = \alpha(\Psi(x))t + \gamma(\Psi(x))\frac{||x||_H^2}{2} + \Phi(x), \tag{7.39}$$

where

$$\varphi(\xi) = \int \frac{d\xi}{\beta(\xi)}, \quad \gamma(a) = a\alpha(a)$$

and $\Psi(x)$, $\Phi(x)$ are arbitrary harmonic functions on H.
2. Let the function $f(\xi, \eta, \zeta)$ satisfy in addition the following conditions.

(3) The equation

$$f\left(V(\tau, x), \frac{1}{a}\frac{\partial V(\tau, x)}{\partial \tau}, \frac{\partial V(\tau, x)}{\partial \tau}\right) = 0 \tag{7.40}$$

can be solved with respect to a, $a = \sigma(V(\tau, x), \partial V(\tau, x)/\partial \tau)$.

Assume that the equation

$$\alpha_a'\left(\sigma\left(V(\tau, x), \frac{\partial V(\tau, x)}{\partial \tau}\right)\Big|_{\tau=X-\frac{\|x\|_H^2}{2}}\right)t$$

$$+ \gamma_a'\left(\sigma\left(V(\tau, x), \frac{\partial V(\tau, x)}{\partial \tau}\right)\Big|_{\tau=X-\frac{\|x\|_H^2}{2}}\right)\left[\frac{\|x\|_H^2}{2} - X\right] = 0, \quad (7.41)$$

can be solved with respect to $X = \chi(t, x)$, with $\chi(0, x) = \|x\|_H^2/2$. Here $V(\tau, x)$ is the solution of the Cauchy problem for the 'heat equation'

$$\frac{\partial V(\tau, x)}{\partial \tau} = \Delta_L V(\tau, x), \quad V(0, x) = U_0(x). \quad (7.42)$$

Let the function φ exist and have an inverse φ^{-1}.

Then the solution of the Cauchy problem

$$f\left(U(t, x), \frac{\partial U(t, x)}{\partial t}, \Delta_L U(t, x)\right) = 0, \quad U(0, x) = U_0(x)$$

for the nonlinear equation (7.37) has the form

$$\varphi(U(t, x)) = \alpha(\Psi(t, x))t + \gamma(\Psi(t, x))\frac{\|x\|_H^2}{2} + \Phi(t, x), \quad (7.43)$$

where

$$\varphi(\xi) = \int \frac{d\xi}{\beta(\xi)}, \quad \gamma(a) = a\alpha(a),$$

$$\Psi(t, x) = \sigma\left(V(\tau, x), \frac{\partial V(\tau, x)}{\partial \tau}\right)\Big|_{\tau=\chi(t,x)-\frac{\|x\|_H^2}{2}},$$

$$\Phi(t, x) = \varphi\left(V\left(\chi(t, x) - \frac{\|x\|_H^2}{2}, x\right)\right) - \gamma(\Psi(t, x))\chi(t, x);$$

$V(t, x)$ is solution of the Cauchy problem (7.42).

If in addition equations (7.38), (7.40), and (7.41) have respectively unique solutions η, a and $\chi(t, x)$ and the Cauchy problem for the 'heat equation' has a unique solution in a certain functional class, then the solution to the Cauchy problem for equation (7.37) is unique in the same class.

Proof. 1. From (7.39), by (1.4) we derive

$$\Delta_L \varphi(U(t, x)) = \varphi_\xi'(U(t, x))\Delta_L U(t, x)$$

$$= \alpha_a'(\Psi(x))\Delta_L \Psi(x)t + \gamma_a'(\Psi(x))\Delta_L \Psi(x)\frac{\|x\|_H^2}{2}$$

$$+ \gamma(\Psi(x))\frac{1}{2}\Delta_L\|x\|_H^2 + \Delta_L\Phi(x)$$

$$= \Psi(x)\alpha(\Psi(x))$$

(since $\Delta_L \Psi(x) = \Delta_L \Phi(x) = 0$, $\Delta_L \|x\|_H^2 = 2$).

Moreover,

$$\frac{\partial \varphi(U(t,x))}{\partial t} = \varphi'_\xi(U(t,x)) \frac{\partial U(t,x)}{\partial t} = \alpha(\Psi(x)).$$

From this we deduce that

$$\frac{\partial U(t,x))}{\partial t} = \alpha(\Psi(x))\beta(U(t,x)) = \phi(U(t,x), \Psi(x)),$$

$$\Delta_L U(t,x) = \Psi(x)\alpha(\Psi(x))\beta(U(t,x)) = \Psi(x)\phi(U(t,x), \Psi(x))$$

since $\varphi'_\xi(\xi) = 1/\beta(\xi)$.

Substituting these relations into (7.37) we derive

$$f\big(U(t,x), \phi(U(t,x), \Psi(x)), \Psi(x)\phi(U(t,x), \Psi(x))\big) = 0.$$

The last equality holds identically as, due to condition (2), $\phi(U(t,x), \Psi(x)) = \phi(U(t,x), \Psi(x))$.

2. Let us rewrite (7.43) in the form

$$\varphi(U(t,x)) = \alpha(\Psi(t,x)t + \gamma(\Psi(t,x))\left[\frac{\|x\|_H^2}{2} - \chi(t,x)\right]$$

$$+ \varphi\left(V\left(\chi(t,x) - \frac{\|x\|_H^2}{2}, x\right)\right). \tag{7.44}$$

From (7.44)

$$\frac{\partial \varphi(U(t,x))}{\partial t} = \varphi'_\xi(U(t,x)) \frac{\partial U(t,x)}{\partial t}$$

$$= \alpha(\Psi(t,x))$$

$$+ \left\{\alpha'_a(\Psi(t,x))t + \gamma'_a(\Psi(t,x))\left[\frac{\|x\|_H^2}{2} - \chi(t,x)\right]\right\}\frac{\partial \Psi(t,x)}{\partial t}$$

$$- \left[\gamma(\Psi(t,x)) - \varphi'_\xi(V(\tau,x))\frac{\partial V(\tau,x)}{\partial \tau}\Big|_{\tau=\chi(t,x)-\frac{\|x\|_H^2}{2}}\right]\frac{\partial \chi(t,x)}{\partial t}. \tag{7.45}$$

Since $\Psi(t,x) = \sigma(V, (\partial V/\partial \tau))|_{\tau=\chi(t,x)+(\|x\|_H^2/2)}$ and $\chi(t,x)$ solve (7.41), which yields $\alpha'_a(\sigma)t + \gamma'_a(\sigma)[(\|x\|_H^2/2) - \chi(t,x)] = 0$, we deduce from (7.45)

$$\varphi'_\xi(U(t,x))\frac{\partial U(t,x)}{\partial t} = \alpha(\Psi(t,x))$$

$$- \left[\gamma(\Psi(t,x)) - \varphi'_\xi(V(\tau,x))\frac{\partial V(\tau,x)}{\partial \tau}\Big|_{\tau=\chi(t,x)-\frac{\|x\|_H^2}{2}}\right]\frac{\partial \chi(t,x)}{\partial t}. \tag{7.46}$$

Furthermore, $\Psi = \sigma(V, \partial V/\partial \tau)$, and hence by condition (3) satisfies the equation

$$f\left(V, \frac{1}{\Psi}\frac{\partial V}{\partial \tau}, \frac{\partial V}{\partial \tau}\right) = 0.$$

Thus, since $f(V, (1/\Psi)(\partial V/\partial \tau), \Psi(1/\Psi)(\partial V/\partial \tau)) = 0$ and by condition (2) we get

$$\frac{1}{\Psi}\frac{\partial V}{\partial \tau} = \phi(V, \Psi).$$

Taking into account that $\gamma(\Psi) = \Psi\alpha(\Psi)$ and $\varphi'_\xi(\xi) = 1/\beta(\xi)$ we derive

$$\gamma(\Psi) - \frac{1}{\beta(V)} \cdot \frac{\partial V}{\partial \tau} = \frac{\alpha(\Psi)\beta(V) - \frac{1}{\Psi} \cdot \frac{\partial V}{\partial \tau}}{\beta(V)} = \frac{\phi(V, \Psi) - \frac{1}{\Psi} \cdot \frac{\partial V}{\partial t}}{\beta(V)} = 0$$

and

$$\gamma(\Psi(t, x)) = \frac{1}{\beta(V(\tau, x))} \cdot \frac{\partial V(\tau, x)}{\partial \tau}\bigg|_{\tau = \chi(t,x) - \frac{\|x\|_H^2}{2}}. \tag{7.47}$$

Substituting (7.47) into (7.46) we obtain

$$\frac{\partial U(t, x)}{\partial t} = \phi(U(t, x), \Psi(t, x)). \tag{7.48}$$

From (7.44), by (1.4) we derive

$$\Delta_L \varphi(U(t, x)) = \varphi'_\xi(U(t, x))\Delta_L U(t, x)$$
$$= \gamma(\Psi(t, x))$$
$$+ \left\{\alpha'_a(\Psi(t, x))t + \gamma'_a(\Psi(t, x))\left[\frac{\|x\|_H^2}{2} - \chi(t, x)\right]\right\}\Delta_L\Psi(t, x)$$
$$- \gamma(\Psi(t, x))\Delta_L\chi(t, x)$$
$$+ \varphi'_\xi(V(\tau, x))\bigg|_{\tau = \chi(t,x) - \frac{\|x\|_H^2}{2}}\Delta_L\left(V(\chi(t, x) - \frac{\|x\|_H^2}{2}, x\right). \tag{7.49}$$

Since $\Psi(t, x) = \sigma(V, (\partial V/\partial \tau))|_{\tau = \chi(t,x) - \frac{\|x\|^2}{2}}$, and $\chi(t, x)$ solves (7.41) we get from (7.49) that

$$\varphi'_\xi(U(t, x))\Delta_L U(t, x) = \gamma(\Psi(t, x)) - \gamma(\Psi(t, x))\Delta_L\chi(t, x)$$
$$+ \varphi'_\xi(V(\tau, x))\bigg|_{\tau = \chi(t,x) - \frac{\|x\|_H^2}{2}}\Delta_L V\left(\chi(t, x) - \frac{\|x\|_H^2}{2}, x\right). \tag{7.50}$$

But

$$\Delta_L V\left(\chi(t, x) - \frac{\|x\|_H^2}{2}, x\right)$$
$$= \frac{\partial V(\tau, x)}{\partial \tau}\bigg|_{\tau = \chi(t,x) - \frac{\|x\|_H^2}{2}}$$
$$\times \left[\Delta_L\chi(t, x) - \frac{1}{2}\Delta_L\|x\|_H^2\right] + \Delta_L V(\tau, x)\bigg|_{\tau = \chi(t,x) - \frac{\|x\|_H^2}{2}}$$

$$= \frac{\partial V(\tau, x)}{\partial \tau}\bigg|_{\tau = \chi(t,x) - \frac{\|x\|_H^2}{2}} [\Delta_L \chi(t, x) - 1] + \Delta_L V(\tau, x)\bigg|_{\tau = \chi(t,x) - \frac{\|x\|_H^2}{2}}$$

$$= \frac{\partial V(\tau, x)}{\partial \tau}\bigg|_{\tau = \chi(t,x) - \frac{\|x\|_H^2}{2}} \Delta_L \chi(t, x)$$

$$- \left[\frac{\partial V(\tau, x)}{\partial \tau} - \Delta_L V(\tau, x)\right]\bigg|_{\tau = \chi(t,x) - \frac{\|x\|_H^2}{2}}$$

$$= \frac{\partial V(\tau, x)}{\partial \tau}\bigg|_{\tau = \chi(t,x) - \frac{\|x\|_H^2}{2}} \Delta_L \chi(t, x)$$

since, by the condition of the theorem, $\partial V(\tau, x)/\partial \tau = \Delta_L V(\tau, x)$.

If we substitute the value of $\Delta_L V(\chi(t, x) - (\|x\|_H^2/2), x)$ in (7.50) we obtain

$$\varphi_\xi'(U(t, x))\Delta_L U(t, x)$$

$$= \gamma(\Psi(t, x)) - \left[\gamma(\Psi(t, x)) - \varphi_\xi'(V(\tau, x))\frac{\partial V(\tau, x)}{\partial \tau}\bigg|_{\tau = \chi(t,x) - \frac{\|x\|_H^2}{2}}\right]\Delta_L \chi(t, x).$$

Now keeping in mind (7.47) we derive

$$\varphi_\xi'(U(t, x))\Delta_L U(t, x) = \gamma(\Psi(t, x))$$

and

$$\Delta_L U(t, x) = \Psi(t, x)\phi(U(t, x), \Psi(t, x)). \tag{7.51}$$

Substituting (7.48) and (7.51) into (7.37) we get

$$f\big(U(t, x), \phi(U(t, x), \Psi(t, x)), \Psi(t, x)\phi(U(t, x), \Psi(t, x))\big) = 0.$$

By condition (2) of the theorem this equality holds identically.

If we put $t = 0$ in (7.43), take into account that $\chi(0, x) = (1/2)\|x\|_H^2$ and consequently $\Phi(0, x) = \varphi(V(0, x)) - \gamma(\Psi(0, x))\|x\|_H^2/2$, we obtain

$$\varphi(U(0, x)) = \gamma(\Psi(0, x))\frac{\|x\|_H^2}{2} + \varphi(V(0, x)) - \gamma(\Psi(0, x))\frac{\|x\|_H^2}{2}$$

$$= \varphi(U_0(x))$$

which yields $U(0, x) = U_0(x)$.

The final assertion of the theorem is obvious. □

Example 7.6 Let us solve the Cauchy problem

$$\left(\frac{\partial U(t, x)}{\partial t}\right)^3 - U(t, x)\left(\frac{\partial U(t, x)}{\partial t}\right)^2 + \left(\frac{\partial U(t, x)}{\partial t}\right)\Delta_L U(t, x)$$

$$= U(t, x)\Delta_L U(t, x), \quad U(0, x) = \|x\|_H^4. \tag{7.52}$$

In this case we have

$$f(\xi, \eta, \zeta) = \eta^3 - \xi\eta^2 + \eta\zeta = \xi\zeta$$

and (7.38) has the form

$$\eta^3 - \xi\eta^2 + a\eta^2 = a\xi\eta.$$

This equation has the non-trivial solutions $\eta = -a$ and $\eta = \xi$.

Consider first the solution to (7.38) of the form $\eta = \phi(\xi, a) = -a$. Here $\alpha(a) = -a$, $\gamma(a) = -a^2$, $\beta(\xi) = 1$, $\varphi(\xi) = \xi$.

Rewriting (7.40) in the form

$$\frac{(\frac{\partial V}{\partial \tau})^3}{a^3} - V\frac{(\frac{\partial V}{\partial \tau})^2}{a^2} + \frac{(\frac{\partial V}{\partial \tau})^2}{a} = V\frac{\partial V}{\partial \tau}$$

we get

$$a = \sigma\left(V, \frac{\partial V}{\partial \tau}\right) = \frac{\partial V}{\partial \tau} \Big/ V.$$

The solution to the Cauchy problem

$$\frac{\partial V(\tau, x)}{\partial \tau} = \Delta_L V(\tau, x), \quad V(0, x) = \|x\|_H^4$$

is given by (see Section 5.4)

$$V(\tau, x) = 4\left[\tau + \frac{\|x\|_H^2}{2}\right]^2$$

and hence

$$\sigma\left(V(\tau, x), \frac{\partial V(\tau, x)}{\partial \tau}\right)\Big|_{\tau = X - \frac{\|x\|_H^2}{2}} = \frac{2}{X}.$$

Equation (7.41) is now of the form

$$(t - 4)X + 2\|x\|_H^2 = 0$$

and has the solution

$$X = \chi(t, x) = -\frac{2\|x\|_H^2}{t - 4}.$$

From (7.43) we derive the solution of the problem (7.52):

$$U(t, x) = \frac{t(t - 4)^3 + 32\|x\|_H^6}{2(t - 4)^2\|x\|_H^2}.$$

This solution is defined for all points $(t, x) \in [0, T] \times H$ except $(4, x)$ or $(t, 0)$.

Consider now another solution $\eta = \phi(\xi, a) = \xi$ to (7.38). Here $\alpha(a) = 1$, $\gamma(a) = a$, $\beta(\xi) = \xi$, $\varphi(\xi) = \ln \xi$. This time (7.41) has the form

$$X - \frac{\|x\|_H^2}{2} = 0,$$

and its solution is given by

$$X = \chi(t, x) = \frac{\|x\|_H^2}{2}.$$

From (7.43) we derive one more solution of the problem (7.52):

$$U(t, x) = \|x\|_H^4 e^t.$$

This solution is defined for all points $t \in [0, T]$, $x \in H$.

Remark. In the particular case when (7.37) can be solved with respect to $\partial U(t, x)/\partial t$ and has the form

$$\frac{\partial U(t, x)}{\partial t} = f_0(U(t, x), \Delta_L U(t, x)),$$

where $f_0(\xi, \zeta)$ is a function of two arguments, the function $\sigma(V, \partial V/\partial \tau)$ from (7.41) is determined by the data of the problem. Indeed, in this case (7.40) is written in the form

$$\frac{1}{a} \frac{\partial V}{\partial \tau} = f_0\left(V, \frac{\partial V}{\partial \tau}\right)$$

which yields

$$a = \sigma\left(V, \frac{\partial V}{\partial \tau}\right) = \frac{\frac{\partial V}{\partial \tau}}{f_0\left(V, \frac{\partial V}{\partial \tau}\right)}.$$

Note that the Cauchy problem for the equation $\partial U(t, x)/\partial t = f_0(U(t, x), \Delta_L U(t, x))$ was considered in Section 7.2.

Appendix

Lévy–Dirichlet forms and associated
Markov processes

Dirichlet forms appear in various problems of modern mathematical physics, stochastic analysis, quantum mechanics and many other areas. There is a close connection between these forms and Markov processes (see, e.g. [7]).

We construct a Dirichlet form associated with the infinite-dimensional Lévy–Laplace operator and a Markov process generated by this Dirichlet form.

To this end we use the self-adjoint operator in $\mathcal{L}_2(H, \mu)$ generated by the non-symmetrized Lévy Laplacian from Section 3.3.

A.1 The Dirichlet forms associated with the
Lévy–Laplace operator

Recall that a Dirichlet form is by definition a Markovian closed symmetric bilinear form.

A symmetric form $\mathcal{E}(U, V)$ on $\mathcal{L}_2(H, \mu)$ is called Markovian if the following property holds [67]: for each $\varepsilon > 0$, there exists a real function $\phi_\varepsilon(t), t \in \mathbb{R}^1$, such that

$$\phi_\varepsilon(t) = t \quad \text{for all } t \in [0, 1], \qquad -\varepsilon \leq \phi_\varepsilon(t) \leq 1 + \varepsilon \quad \text{for all } t \in \mathbb{R}^1,$$

$$0 \leq \phi_\varepsilon(s) - \phi_\varepsilon(t) \leq s - t \quad \text{whenever} \quad t < s,$$

and for any $U \in D_\mathcal{E}$, we have that

$$\phi_\varepsilon(U) \in D_\mathcal{E} \quad \text{and} \quad \mathcal{E}(\phi_\varepsilon(U), \phi_\varepsilon(U)) \leq \mathcal{E}(U, U).$$

Recall the operator $T = K^{-1/2}$ introduced in Section 2.1 (K is the correlation operator of the Gaussian measure μ). Require in addition that $T^{-1/2}$ is a Hilbert–Schmidt operator. Assume and that $\sum_{k=1}^\infty \sqrt{(T^{-2} f_k, f_k)_H} < \infty$. For example, if $\{f_k\}_1^\infty$ coincides with the canonical basis $\{e_k\}_1^\infty$ in H, then

$$\sum_{k=1}^\infty \sqrt{(T^{-2} e_k, e_k)_H} = \sum_{k=1}^\infty \frac{1}{\lambda_k} = \operatorname{Tr} T^{-1} < \infty$$

since T^{-1} is a trace operator.

We consider the Lévy–Laplace operator $LU = \Delta_L U$, $D_L = \mathfrak{T}$ introduced in Section 3.3, where the set $\mathfrak{T}$ consists of functions of the form $V(x) = \varphi(Q(x))S(x)$.

Moreover the function $Q(x)$ is chosen to be of the form $Q(x) = \|x\|_H^2 + \phi(x)$, with $\phi(x) = \sum_{k=1}^{\infty}(x, f_k)_H$. Such a choice of $Q(x)$ makes $D_L \cap D_\mathcal{E} \neq \varnothing$, where $\mathcal{E}$ is the Dirichlet form.

Denote by $\mathfrak{T}$ the set of all functions of the form

$$V(x) = \varphi(Q(x))S(x)$$

such that $V(x), [\varphi'(Q(x))/\varphi(Q(x))]^2 V(x) \in \mathfrak{L}_2(H, \mu)$. Here $S(x)$ is an arbitrary, twice strongly differentiable harmonic in H functions from $\mathfrak{L}_2(H, \mu)$,

$$Q(x) = \sum_{k=1}^{\infty} \zeta_k(x) = \|x\|_H^2 + \phi(x), \tag{A.1}$$

$$\zeta_k(x) = (x, f_k)_H^2 + (x, f_k)_H, \phi(x) = \sum_{k=1}^{\infty}(x, f_k)_H(f_k \in H_+, x \in H),$$

and $\varphi(\xi)$ is a fixed function on R^1.

We choose $\varphi(\xi)$ to be a solution to the equation

$$\varphi(\xi)\varphi''(\xi) + 2\varphi(\xi)\varphi'(\xi) + [\varphi'(\xi)]^2 = 0. \tag{A.2}$$

By the substitution $\kappa(\xi) = \varphi'(\xi)/\varphi(\xi)$, for ξ such that $\varphi(\xi) \neq 0$, (A.2) can be reduced to the Riccati equation

$$\kappa'(\xi) + 2\kappa^2(\xi) + 2\kappa(\xi) = 0. \tag{A.3}$$

The series $\sum_{k=1}^{\infty} \zeta_k(x)$, converges μ-almost everywhere on H, since $\phi(x) < \infty$, for μ-almost all $x \in H$. Actually

$$\int_H |(x, f_k)_H|\mu(dx) = \frac{2\sqrt{(Kf_k, f_k)_H}}{\sqrt{2\pi}} \int_0^{\infty} ye^{-\frac{1}{2}y^2}dy = \sqrt{\frac{2}{\pi}(Kf_k, f_k)_H}.$$

Hence $\sum_{k=1}^{\infty} \int_H |(x, f_k)_H|\mu(dx) < \infty$ and by Levi's theorem $\sum_{k=1}^{\infty}(x, f_k)_H < \infty$, μ-almost everywhere on H.

The function $V(x) = \varphi(Q(x))S(x)$ is twice differentiable along the subspace H_+ since both $\phi(x) = \sum_{k=1}^{\infty}(x, f_k)_H$ and $\|x\|_H^2$ and hence the above $Q(x)$ given by (A.1) possess this property. To check that $\phi(x)$ is twice differentiable let us compute

$$d\phi(x; h) = \sum_{k=1}^{\infty}(h, f_k)_H \leq \sum_{k=1}^{\infty}[(Th, Th)_H(T^{-1}f_k, T^{-1}f_k)_H]^{\frac{1}{2}}$$

$$= \|h\|_{H_+} \sum_{k=1}^{\infty}\sqrt{(T^{-2}f_k, f_k)_H} < \infty, \quad d^2\phi(x; h) = 0, \quad (h \in H_+).$$

It is obvious that the set $\mathfrak{T}$ is linear and is everywhere dense in $\mathfrak{L}_2(H, \mu)$.

Define on $\mathfrak{L}_2(H, \mu)$ the bilinear form $\mathcal{E}(U, V)$ by

$$\mathcal{E}(U, V) = \lim_{n \to \infty} \mathcal{E}_n(U, V),$$

where

$$\mathcal{E}_n(U, V) = \int_H \frac{1}{n} \sum_{k=1}^{n}(U'_{H_+}(x), f_k)_H(V'_{H_+}(x), f_k)_H\mu(dx).$$

Lemma A.1 *The form $\mathcal{E}(U, V)$ on $\mathfrak{T}$ exists and*

$$\mathcal{E}(U, V) = (\kappa^2(Q)U, V)_{\mathfrak{L}_2(H, \mu)}$$

where $\kappa(\xi)$ is a positive solution of the Riccati equation (A.3). The form $\mathcal{E}(U, V)$ is symmetric and densely defined.

Proof. For $U, V \in \mathfrak{T}$ we have $U(x) = \varphi(Q(x))S_U(x)$, $V(x) = \varphi(Q(x))S_V(x)$. Therefore

$$dU(x; h) = (U'_{H_+}(x), h)_H$$
$$= \varphi'(Q(x))(Q'(x), h)_H S_U(x) + \varphi(Q(x))(S'_U(x), h)_H,$$
$$dV(x; h) = (V'_{H_+}(x), h)_H$$
$$= \varphi'(Q(x))(Q'(x), h)_H S_V(x) + \varphi(Q(x))(S'_V(x), h)_H \quad (h \in H_+),$$

and

$$\frac{1}{n} \sum_{k=1}^{n} (U'_{H_+}(x), f_k)_H (V'_{H_+}(x), f_k)_H$$

$$= [\varphi'(Q(x))]^2 S_U(x)S_V(x)\frac{1}{n} \sum_{k=1}^{n}(Q'(x), f_k)^2_H$$

$$+ \varphi'(Q(x))\varphi(Q(x))\frac{1}{n} \sum_{k=1}^{m}(Q'(x), f_k)_H [S_U(x)(S'_V(x), f_k)_H$$

$$+ S_V(x)(S'_U(x), f_k)_H] + \varphi^2(Q(x))\frac{1}{n} \sum_{k=1}^{m}(S'_U(x), f_k)_H(S'_V(x), f_k)_H.$$

Since

$$dQ(x; h) = (Q'(x), h)_H = \sum_{k=1}^{\infty} 2(x, f_k)_H (f_k, h)_H + (h, f_k)_H,$$

we obtain $(Q'(x), f_k)_H = 2(x, f_k)_H + (f_k, f_k)_H$. Furthermore

$$\lim_{n\to\infty} \frac{1}{n} \sum_{k=1}^{n}(Q'(x), f_k)^2_H = 1$$

due to the convergence $(x, f_k)_H \to 0$ as $k \to \infty$, μ-almost all $x \in H$. Finally we have

$$\lim_{n\to\infty} \frac{1}{n} \sum_{k=1}^{n}(U'_{H_+}(x), f_k)_H(V'_{H_+}(x), f_k)_H = [\varphi'(Q(x))]^2 S_U(x)S_V(x)$$

$$= \left[\frac{\varphi'(Q(x))}{\varphi(Q(x))}\right]^2 U(x)V(x) = \kappa^2(Q(x))U(x)V(x)$$

and

$$\mathcal{E}(U, V) = (\kappa^2(Q)U, V)_{\mathfrak{L}_2(H, \mu)} \quad U, V \in \mathfrak{T}.$$

$\square$

Lemma A.2 *The Lévy Laplacian Δ_L exists on $\mathfrak{T}$ and acts as the operator of multiplication by the function $-\kappa^2(Q(x))$*

$$\Delta_L V(x) = -\kappa^2(Q(x))V(x),$$

where $\kappa(\xi)$ is a positive solution of the Riccati equation (A.3).

Proof. For $V(x) \in \mathfrak{T}$, we have $V(x) = \varphi(Q(x))S_V(x)$. Therefore

$$dV(x;h) = (V'_{H_+}(x), h)_H$$
$$= \varphi'(Q(x))(Q'(x), h)_H S_V(x) + \varphi(Q(x))(S'_V(x), h)_H,$$
$$d^2V(x;h) = (V''_{H_+}(x)h, h)_H$$
$$= \varphi''(Q(x))(Q'(x), h)^2_H S_V(x)$$
$$+ \varphi'(Q(x))(Q''(x)h, h)_H S_V(x) + 2\varphi'(Q(x))(Q'(x), h)_H(S'_V(x), h)_H$$
$$+ \varphi(Q(x))(S''_V(x)h, h)_H.$$

From the formula (1.3)

$$\Delta_L V(x) = \varphi''(Q(x))S_V(x) \lim_{n\to\infty} \frac{1}{n} \sum_{k=1}^n (Q'(x, f_k)^2_H$$

$$+ \varphi'(Q(x))S_V(x) \lim_{n\to\infty} \frac{1}{n} \sum_{k=1}^n (Q''(x)f_k, f_k)_H$$

$$+ 2\varphi'(Q(x)) \lim_{n\to\infty} \frac{1}{n} \sum_{k=1}^n (Q'(x), f_k)_H(S'_V(x), f_k)_H$$

$$+ \varphi(Q(x)) \lim_{n\to\infty} \frac{1}{n} \sum_{k=1}^n (S''_V(x)f_k, f_k)_H$$

$$= \varphi''(Q(x))S_V(x) \lim_{n\to\infty} \frac{1}{n} \sum_{k=1}^m (Q'(x), f_k)^2_H$$

$$+ \varphi'(Q(x))S_V(x) \lim_{n\to\infty} \frac{1}{n} \sum_{k=1}^n (Q''(x)f_k, f_k)_H,$$

since S_V is a harmonic function.

In the proof of Lemma A.1 we have shown that for μ-almost all $x \in H$

$$\lim_{n\to\infty} \frac{1}{n} \sum_{k=1}^n (Q'(x), f_k)^2_H = 1.$$

In addition since $d^2 Q(x; h) = (Q''(x)h, h)_H = 2\|h\|^2_H$, we have

$$(Q''(x)f_k, f_k)_H = 2,$$

and hence

$$\Delta_L Q(x) = \lim_{n\to\infty} \frac{1}{n} \sum_{k=1}^n (Q''(x)f_k, f_k) = 2.$$

Hence

$$\Delta_L V(x) = \varphi''(Q(x))S_V(x) + 2\varphi'(Q(x))S_V(x) = \frac{\varphi''(Q(x)) + 2\varphi'(Q(x))}{\varphi(Q(x))}V(x).$$

Taking into account that $\varphi(\xi)$ is governed by (A.2) we get

$$\Delta_L V(x) = -\left[\frac{\varphi'(Q(x))}{\varphi(Q(x))}\right]^2 V(x) = -\kappa^2(Q(x))V(x).$$

□

Define the operator L in $\mathfrak{L}_2(H, \mu)$ with everywhere dense domain of definition D_L putting

$$LU = \Delta_L U, \quad D_L = \mathfrak{T}.$$

Theorem A.1 *The operator L is an essentially self-adjoint operator and $-L$ is positive. The closure $-\bar{L}$ of $-L$ is a self-adjoint positive operator.*

Proof. The proof of essential self-adjointness of L is similar to the proof in Theorem 3.2. Furthermore, since a general solution $\kappa(\xi)$ to the Riccati equation (A.3) has the form

$$\kappa(\xi) = -\frac{1}{1 + ce^{2\xi}},$$

where c is a constant and $\kappa^2(\xi) > 0$, for $-\infty < \xi < \infty$ we obtain

$$(-LU, U)_{\mathfrak{L}_2(H,\mu)} = (\kappa^2(Q)U, U)_{\mathfrak{L}_2(H,\mu)} > 0, \qquad \text{for all } U \in D_L.$$

The final conclusions of the theorem are evident.

□

Consider the form

$$\mathcal{E}(U, V) = \left(\sqrt{-\bar{L}}U, \sqrt{-\bar{L}}V\right)_{\mathfrak{L}_2(H,\mu)} \qquad U, V \in D_{\mathcal{E}}, \quad D_{\mathcal{E}} = D_{\sqrt{-\bar{L}}}. \tag{A.4}$$

Theorem A.2 *The form $\mathcal{E}(U, V)$ given by (A.4) is symmetric, densely defined, positive and closed. It is a Dirichlet form in $\mathfrak{L}_2(H, \mu)$: that is $\mathcal{E}(U, V)$ is a symmetric, closed, bilinear Markov form.*

Proof. The first three assertions result from Lemma A.1, Lemma A.2 and Theorem A.1. Furthermore the form $\mathcal{E}(U, V)$ is closed since $\sqrt{-\bar{L}}$ is a closed operator (recall that $\sqrt{-\bar{L}}$ is positive and self-adjoint).

To prove that $\mathcal{E}(U, V)$ is a Dirichlet form we use the sufficient condition from [67]. Let

$$U \in D_{\mathcal{E}}, \quad V = (0 \vee U) \wedge 1.$$

Then $V \in D_{\mathcal{E}}$ and estimate $\mathcal{E}(V, V) \le \mathcal{E}(U, U)$ is obvious since $\mathcal{E}(U, V) = (|\kappa(Q)|U, |\kappa(Q)|V)_{\mathfrak{L}_2(H,\mu)}$.

□

Due to Theorem A.2 it is natural to call $\mathcal{E}(U, V)$ the Lévy–Dirichlet form.

A.2 The stochastic processes associated with the Lévy–Dirichlet forms

Theorem A.3 *A Markov process $\xi_x(t)$ on H, associated with the Lévy–Dirichlet form generated by the Lévy–Laplace operator $\bar{L}$, can be constructed as the limit (for $n \to \infty$)*

of a family of the diffusion processes $\xi_{x,n}(t)$ associated with the forms

$$\mathcal{E}_n(U, V) = (\sqrt{l_n}\, U, \sqrt{l_n}\, V)_{\mathcal{L}_2(H, \mu)}, \tag{A.5}$$

where

$$l_n U(x) = -\frac{1}{n} \sum_{k=1}^{n} \left[(U''_{H_+}(x) f_k, f_k)_H - (U'_{H_+}(x), f_k)_H(x, f_k)_{H_+} \right]. \tag{A.6}$$

Proof. The forms $\mathcal{E}_n(U, V)$ are symmetric and densely defined. Applying the integration by parts formula (2.2) for $n = 1$ and for $F = VdU$, we rewrite (A.5) in the form

$$\mathcal{E}_n(U, V) = \int_H \frac{1}{n} \sum_{k=1}^{n} (U'_{H_+}(x), f_k)_H (V'_{H_+}(x), f_k)_H \mu(dx) \quad U, V \in D_{\mathcal{E}}$$

(see Lemma A.1).

Let $f_k = e_k$, where $\{e_k\}_1^\infty$ is the canonical basis in H. Since

$$U'_{H_+}(x) = \sum_{k=1}^{\infty} \frac{\partial U}{\partial x_k} e_k, \quad U''_{H_+}(x) e_j = \sum_{k=1}^{\infty} \frac{\partial^2 U}{\partial x_j \partial x_k} e_k,$$

$$x_k = (x, e_k)_H, \quad (x, e_k)_{H_+} = \lambda_k^2 x_k,$$

we deduce from (A.6) that

$$l_n = -\frac{1}{n} \sum_{k=1}^{n} \left[\frac{\partial^2}{\partial x_k^2} - \lambda_k^2 x_k \frac{\partial}{\partial x_k} \right]. \tag{A.7}$$

For each $n \in N$ the symmetrized n-dimensional Laplacian l_n is positive and essentially self-adjoint on the dense set $\mathfrak{T}$ in $\mathcal{L}_2(H, \mu)$. It generates a contractive positivity preserving semigroup

$$\mathcal{T}_n(t) = e^{-l_n t} \quad (t \geq 0)$$

on $\mathcal{L}_2(H, \mu)$ such that $\mathcal{T}_n(t) \cdot 1 = 1$. Hence $\mathcal{T}_n(t)$ is a Markov semigroup and thus the corresponding form $\mathcal{E}_n(U, V)$ is a Markov form.

There is one-to-one correspondence between the semigroup $\mathcal{T}_n(t)$ and the transition probability $P_n(t, x, B) = P\{\xi_{x,n}(t) \in B | \xi_{x,n}(0) = x\}$ of a diffusion process $\xi_{x,n}(t)$ defined on the probability space $(\Omega, \mathcal{F}, P)$; B is a Borel subset of H.

For any bounded measurable function f one has

$$(\mathcal{T}_n(t) f)(x) = \int_H f(y) P(t, x, dy) = E(f(\xi_{x,n}(t))) \tag{A.8}$$

for μ-almost all $x \in H$ and $t \geq 0$. For any bounded continuous function f on H there exists a unique solution to the Cauchy problem $\partial U(t, x)/\partial t + l_n U(t, x) = 0$, for $t > 0$, with $U(0, x) = f(x)$ considering $U(t, x)$ as a function of $x_1, \ldots, x_n$, the variables $x_{n+1}, x_{n+2}, \ldots$ of $U(t, x)$ being considered as parameters.

We prove that $\mathcal{T}_n(t)$ converges strongly in $\mathcal{L}_2(H, \mu)$ to $\mathcal{T}(t)$, that is for all $f \in \mathcal{L}_2(H, \mu) \|\mathcal{T}_n(t) f - \mathcal{T}(t) f\|_{\mathcal{L}_2(H, \mu)} \to 0$ as $n \to \infty$.

Choose $f \in D_{\mathcal{E}}$. Then

$$\|e^{-t\bar{l}_m}f - e^{-t\bar{l}_n}f\|_{\mathfrak{L}_2(H,\mu)} \leq \|e^{-t\bar{l}_m}f - e^{-t\frac{n}{m}\bar{l}_n}f\|_{\mathfrak{L}_2(H,\mu)} + \|e^{-t\frac{n}{m}\bar{l}_n}f - e^{-t\bar{l}_n}f\|_{\mathfrak{L}_2(H,\mu)}.$$

For $m \geq n$ we have $l_m \geq (n/m)l_n$, since

$$(l_m f, f)_{\mathfrak{L}_2(H,\mu)} = \int_H \frac{1}{m} \sum_{k=1}^{m} \left(\frac{\partial f}{\partial x_k}\right)^2 \mu(dx)$$

$$\geq \int_H \frac{1}{m} \sum_{k=1}^{n} \left(\frac{\partial f}{\partial x_k}\right)^2 \mu(dx) = \frac{n}{m}(l_n f, f)_{\mathfrak{L}_2(H,\mu)}.$$

Hence,

$$e^{-tl_m} \leq e^{-t\frac{n}{m}l_n}$$

and

$$\|e^{-t\frac{n}{m}\bar{l}_n}f - e^{-t\bar{l}_m}f\|_{\mathfrak{L}_2(H,\mu)} \leq \sqrt{\|e^{-t\frac{n}{m}\bar{l}_n} - e^{-t\bar{l}_m}\|}\sqrt{\left(\left[e^{-t\frac{n}{m}\bar{l}_n} - e^{-t\bar{l}_m}\right]f, f\right)_{\mathfrak{L}_2(H,\mu)}}$$

$$\leq \sqrt{2\left(\left[e^{-t\frac{n}{m}\bar{l}_n} - e^{-t\bar{l}_m}\right]f, f\right)_{\mathfrak{L}_2(H,\mu)}},$$

since $\|\mathcal{T}_n(t)\| \leq 1$.

By Duhamel's formula we have

$$e^{-t\frac{n}{m}\bar{l}_n}f - e^{-t\bar{l}_m}f = \int_0^t e^{-(t-s)\frac{n}{m}\bar{l}_n}\left[\bar{l}_m - \frac{n}{m}\bar{l}_n\right]e^{-s\bar{l}_m}f\,ds. \tag{A.9}$$

It is easy to check by direct computation that the operators $e^{-(t-s)\frac{n}{m}l_n}$ and e^{-sl_m} commute, since $(n/m)l_n$ and l_m commute. To prove the latter property it is sufficient to recall (A.7). In fact, setting $q_k = (\partial^2/\partial x_k^2) - \lambda_k^2 x_k(\partial/\partial x_k)$ we see immediately that $q_j q_k = q_k q_j$ for all $k, j = 1, \ldots n$ which yields $l_n l_m = l_m l_n$ and thus $(n/m)l_n$ and l_m commute.

From this we get from (A.9)

$$e^{-t\frac{n}{m}\bar{l}_n}f - e^{-t\bar{l}_m}f = \int_0^t e^{-(t-s)\frac{n}{m}\bar{l}_n - s\bar{l}_m}\left[\bar{l}_m - \frac{n}{m}\bar{l}_n\right]f\,ds$$

and

$$\|e^{-t\frac{n}{m}\bar{l}_n}f - e^{-t\bar{l}_m}f\|^2_{\mathfrak{L}_2(H,\mu)} \leq 2\left(\left[e^{-t\frac{n}{m}\bar{l}_n} - e^{-t\bar{l}_m}\right]f, f\right)_{\mathfrak{L}_2(H,\mu)}$$

$$= 2\left(\int_0^t e^{-(t-s)\frac{n}{m}\bar{l}_n - s\bar{l}_m}\left[\bar{l}_m - \frac{n}{m}\bar{l}_n\right]f\,ds, f\right)_{\mathfrak{L}_2(H,\mu)}$$

$$= 2\int_0^t \left(\left[\bar{l}_m - \frac{n}{m}\bar{l}_n\right]f, G_{st}f\right)_{\mathfrak{L}_2(H,\mu)}ds$$

$$\leq 2\sqrt{\left(\left[\bar{l}_m - \frac{n}{m}\bar{l}_n\right]f, f\right)_{\mathfrak{L}_2(H,\mu)}}\int_0^t \sqrt{\left(\left[\bar{l}_m - \frac{n}{m}\bar{l}_n\right]G_{st}f, G_{st}f\right)_{\mathfrak{L}_2(H,\mu)}}\,ds,$$

where $G_{st}f = e^{-(t-s)\frac{n}{m}\bar{l}_n - \bar{l}_m s}\, f$.

But

$$\left\{\int_0^t \sqrt{\left(\left[\bar{l}_m - \frac{n}{m}\bar{l}_n\right] G_{st}f, G_{st}f\right)_{\mathfrak{L}_2(H,\mu)}}\, ds\right\}^2$$

$$\leq t\int_0^t \left(\left[\bar{l}_m - \frac{n}{m}\bar{l}_n\right] G_{st}f, G_{st}f\right)_{\mathfrak{L}_2(H,\mu)} ds$$

$$= t\left(\int_0^t e^{-2(t-s)\frac{n}{m}\bar{l}_n}[\bar{l}_m - \frac{n}{m}\bar{l}_n] e^{-2s\bar{l}_m} f\, ds,\, f\right)_{\mathfrak{L}_2(H,\mu)}$$

$$= t\left(\frac{1}{2}\left[e^{-2t\frac{n}{m}\bar{l}_n} - e^{-2t\bar{l}_m}\right] f,\, f\right)_{\mathfrak{L}_2(H,\mu)}$$

$$\leq \frac{t}{2}\|e^{-2t\frac{n}{m}\bar{l}_n} - e^{-2t\bar{l}_m}\| \|f\|^2_{\mathfrak{L}_2(H,\mu)} \leq t\|f\|^2_{\mathfrak{L}_2(H,\mu)},$$

since $\|T_n(t)\| \leq 1$.

By Lemma A.1 we know that $\mathcal{E}_n(f, f)$ is a Cauchy sequence which implies

$$\|e^{-t\frac{n}{m}\bar{l}_n}f - e^{-t\bar{l}_m}f\|_{\mathfrak{L}_2(H,\mu)} \leq \left\{4t\left(\left[\bar{l}_m - \frac{n}{m}\bar{l}_n\right] f, f\right)_{\mathfrak{L}_2(H,\mu)}\right\}^{1/4}\|f\|^{1/2}_{\mathfrak{L}_2(H,\mu)}$$

$$= \left\{4t\left[\mathcal{E}_m(f, f) - \frac{n}{m}\mathcal{E}_n(f, f)\right]\right\}^{1/4}\|f\|^{1/2}_{\mathfrak{L}_2(H,\mu)}$$

$$\to 0 \text{ as } m, n \to \infty,$$

In addition for $m \geq n$ we have $e^{-t\bar{l}_n} \leq e^{-t\frac{n}{m}\bar{l}_n}$ from which we get

$$\|e^{-t\frac{n}{m}\bar{l}_n}f - e^{-t\bar{l}_n}f\|_{\mathfrak{L}_2(H,\mu)} \leq \left\{4t\left(\left[\bar{l}_n - \frac{n}{m}\bar{l}_n\right] f, f\right)_{\mathfrak{L}_2(H,\mu)}\right\}^{1/4}\|f\|^{1/2}_{\mathfrak{L}_2(H,\mu)}$$

$$= \left\{4t\left(1 - \frac{n}{m}\right)\mathcal{E}_n(f, f)\right\}^{1/4}\|f\|^{1/2}_{\mathfrak{L}_2(H,\mu)}$$

$$\to 0 \text{ as } m, n \to \infty.$$

Thus,

$$\lim_{m>n,n\to\infty} \|e^{-t\bar{l}_m}f - e^{-t\bar{l}_n}f\|_{\mathfrak{L}_2(H,\mu)} = 0 \qquad \text{for all } f \in D_{\mathcal{E}}, \tag{A.10}$$

for any t from a finite interval.

The sequence $e^{-t\bar{l}_n}f$ is a Cauchy sequence for any $t > 0$, $f \in D_{\mathcal{E}}$. Since $D_{\mathcal{E}}$ is dense in $\mathfrak{L}_2(H, \mu)$, and the family $e^{-t\bar{l}_n}$ is uniformly bounded we conclude that (A.10) holds for all $f \in \mathfrak{L}_2(H, \mu)$. Hence

$$\lim_{n\to\infty} T_n(t)f = T(t)f \qquad \text{for all } f \in \mathfrak{L}_2(H, \mu). \tag{A.11}$$

The semigroup $T(t)$ is a contraction in $C_2(H, \mu)$ since in the strong limit this property of $T_n(t)$ is inherited. Moreover $T1 = 1$, since $T_n 1 = 1$ for all n and T is positivity preserving, since the T_n are positivity preserving. Hence T is a Markov semigroup in $\mathcal{L}_2(H, \mu)$. By the Kolmogorov–Ionescu Tulcea construction there exists a Markov process $\xi_x(t)$ such that $E(f(\xi_x(t))) = T(t)f(x)$ for μ-almost all $x \in H$, for any bounded measurable function f defined on H.

From (A.8) and (A.11) we deduce that

$$(T(t)f)(x) = \lim_{n \to \infty} E(f(\xi_{x,n}(t))) = E(f(\xi_x(t)))$$

for μ-almost all $x \in H$.   □

Bibliographic notes

Chapter 1

The Laplacian for functions on a Hilbert space was introduced by Lévy [92–95]. The representation of the Laplacian (1.2) (Lemma 1.1) for the case of functionals defined on functional spaces was obtained by Polishchuk [121]. Formula (1.4) is due to Lévy [92–95]. It was used by many authors. Corollary 3 from this formula is due to Polishchuk in [116]. The theory of Gaussian measures in a Hilbert space is presented in many works (see, e.g., [27, 153]). Measure in the space of continuous functions was defined as early as 1923 by Wiener [156]. In a Hilbert space of functions which are orthogonal to unity, the Wiener measure is introduced in the book by Shilov and Phan Dich Tinh [136].

Chapter 2

This chapter is the extended presentation of the article [50] by Feller (see also [52]). The complete orthonormalized systems of polynomials for the Wiener measure were constructed by Cameron and Martin [25], by Ito [81], by Wiener [157], and for general Gaussian measures – by Vershik [153]. Theorem 2.1 for the case of the Wiener integral of multiple variations was proved by Owchar in [107]. Other theorems of this chapter are due to the author of this book.

Chapter 3

The results of Sections 3.1 and 3.2 were published in the article by Feller [51] (see also [52]), and the results of Section 3.3 were published in the article by Feller [53].

Chapter 4

This chapter presents the results of Feller [54–57, 59].

Chapter 5

5.1. The Dirichlet problem for the Lévy–Laplace and Lévy–Poisson equations was studied by Lévy [92–95], Polishchuk [116, 119], Feller [36, 37], Shilov [132, 134], Dorfman [28], Sikiryavyi [137, 140], Kalinin [82] and Bogdansky [19]. The notion of a fundamental domain in a Hilbert space was introduced by Polishchuk in [116]. The uniqueness theorems were proved by Feller [36, 37].
5.2. The Dirichlet problem for the stationary Lévy–Schrödinger equation in various functional classes was considered by Feller [38] and Shilov [134].
5.3. The Riquier problem for a linear equation in iterated Lévy Laplacians was studied by Polishchuk [117], Shilov [134], and, for a polyharmonic equation, by Feller [40].
5.4. The Cauchy problem for the 'heat equation' in different functional classes was considered in the works of Dorfman [29], Sokolovsky [148], Bogdansky [14, 15, 17, 18, 20] and Bogdansky and Dalecky [23]. The correspondence between the Cauchy problem for the 'heat equation' and the Dirichlet problem for the Lévy–Laplace equation was revealed by Polishchuk in [118].

Theorems 5.1 and 5.2 are due to Shilov [132], Theorems 5.6, 5.7 and 5.13 are due to Polishchuk [116, 117]. Other theorems of this chapter are due to Feller.

Chapter 6

Elliptic quasilinear equations with Lévy Laplacians were studied by Lévy [92, 95], Shilov [134], Sikiryavyi [140], Sokolovsky [146, 150], Feller [61]. Elliptic nonlinear equations with Lévy Laplacians were studied by Feller [60, 62–64].

The presentation of the material in Section 6.1 follows the studies of Lévy [92, 95], and in Sections 6.2– 6.4 we follow articles by Feller [60–64].

Chapter 7

Parabolic nonlinear equations with Lévy Laplacians appear in the works of Shilov [134] (a mixed problem for a nonlinear equation with iterated Lévy Laplacians), Feller [65, 66] (the Cauchy problem for nonlinear parabolic equations). Quasilinear equations appear in the work of Sokolovsky [151].

The presentation of the material in this chapter follows articles by Feller [65, 66].

Appendix

The stochastic processes associated with the Lévy Laplacian were considered by Accardi, Roselli and Smolyanov [5], Accardi and Smolyanov [6], Accardi and Bogachev [1–3], Kuo, Obata and Saito [90] and Albeverio, Belopolskaya and Feller [8].

This Appendix follows the articles by Feller [53] and by Albeverio, Belopolskaya and Feller [8].

References

1. Accardi L. & Bogachev V. The Ornstein–Uhlenbeck process and the Dirichlet form associated to the Lévy Laplacian. *C.R. Acad. Sc. Paris.* 1995, **320**, Serie I, 597–602.
2. Accardi L. & Bogachev V. The Ornstein–Uhlenbeck process associated with the Lévy Laplacian and its Dirichlet form. *Probability and Math. Statistics.* 1997, **17**, Fasc. 1, 95–114.
3. Accardi L. & Bogachev V. On the stochastic analysis associated with the Lévy Laplacian. *Dokl. Akad. Nauk.* 1998, **358**(5), 583–7 (in Russian).
4. Accardi L., Gibilisco P. & Volovich I. Yang–Mills gauge fields and harmonic functions for the Lévy Laplacian. *Russ. J. Math. Phys.* 1994. 2(2), 225–50.
5. Accardi L., Roselli P. & Smolyanov O. G. Brownian motion induced by the Lévy Laplacian. *Mat. Zametki.* 1993, **54**(5), 144–8 (in Russian).
6. Accardi L. & Smolyanov O. G. The Gaussian process induced by the Lévy Laplacian and the Feynman–Kats formula which corresponds to it. *Dokl. Akad. Nauk.* 1995. **342**(4), 442–5 (in Russian).
7. Albeverio S. *Theory of Dirichlet forms and Applications.* Lectures on Probability Theory and Statistics. Lecture Notes in Math. Berlin: Springer, 2003, vol. 1816, pp. 1–106.
8. Albeverio S., Belopolskaya Ya. & Feller M. N. *Lévy Dirichlet forms.* Preprint No. 60, Bonn University, 2003, pp. 1–16.
9. Arnaudon M., Belopolskaya Ya. & Paycha S. Renormalized Laplacians on a class of Hilbert manifolds and a Bochner–Weitzenböck type formula for current groups. *Infinite Dimensional Analysis, Quantum Probability and Related Topics*, 1999, **3**(1), 53–98.
10. Averbukh V. I., Smolyanov O. G. & Fomin S. V. Generalized functions and differential equations in linear spaces. II, Differential operators and Fourier transforms. *Trudy Moskov. Mat. Obshch.* 1972, **27**, 247–82 (in Russian).
11. Berezansky Yu. M. *Expansion in Eigenfunctions of Self-adjoint Operators.* Kiev: Naukova Dumka, 1965 (in Russian).
12. Berezansky Yu. M. & Kondratiev Yu. G. *Spectral Methods in Infinite Dimensional Analysis.* Kiev: Naukova Dumka, 1988 (in Russian).
13. Bogdansky Yu. V. On one class of differential operators of second order for functions of infinite arguments. *Dokl. Akad. Nauk Ukrain. SSR.* 1977, Ser. A, 11, 6–9 (in Russian).

14. Bogdansky Yu. V. The Cauchy problem for parabolic equations with essentially infinite-dimensional elliptic operators. *Ukrain. Mat. Zhurn.* 1977, **29**(6), 781–4 (in Russian).

15. Bogdansky Yu. V. The Cauchy problem for an essentially infinite-dimensional parabolic equation on an infinite-dimensional sphere. *Ukrain. Mat. Zhurn.* 1983, **35**(1), 18–22 (in Russian).

16. Bogdansky Yu. V. Principle of maximum for irregular elliptic differential equation in Hilbert space. *Ukrain. Mat. Zhurn.* 1988. **40**(1), 21–5 (in Russian).

17. Bogdansky Yu. V. Cauchy problem for heat equation with irregular elliptic operator. *Ukrain. Mat. Zhurn.* 1989, **41**(5), 584–90 (in Russian).

18. Bogdansky Yu. V. Cauchy problem for essentially infinite-dimensional parabolic equation with varying coefficients. *Ukrain. Mat. Zhurn.* 1994, **46**(6), 663–70 (in Russian).

19. Bogdansky Yu. V. Dirichlet problem for Poisson equation with essentially infinite-dimensional elliptic operator. *Ukrain. Mat. Zhurn.* 1994, **46**(7), 803–8 (in Russian).

20. Bogdansky Yu. V. Cauchy problem for the essentially infinite-dimensional heat equation on a surface in Hilbert space. *Ukrain. Mat. Zhurn.* 1995, **47**(6), 737–46.

21. Bogdansky Yu. V. The Neuman problem for Laplace equation with the essentially infinite-dimensional elliptic operator. *Dokl. Akad. Nauk Ukrain.* 1995, **10**, 24–6.

22. Bogdansky Yu. V. Essentially infinite-dimensional elliptic operators and P. Lévy's problem. *Methods of Functional Analysis and Topology.* 1999, **5**(4), 28–36.

23. Bogdansky Yu. V. & Dalecky Yu. L. Cauchy problem for the simpliest parabolic equation with essentially infinite-dimensional elliptic operator. Suppl. to chapters IV, V in the book: Dalecky Yu. L. & Fomin S. V. *Measures and Differential Equations in Infinite-dimensional Space.* Amsterdam, New York: Kluwer Acad. Publ., 1991, pp. 309–22.

24. Bogolyubov N. N. On a new method in the theory of superconductivity. III. *Zhurn. Eksp. i Teor. Fiz.* 1958, **34**(1), 73–9 (in Russian).

25. Cameron R. H. & Martin W. T. The orthogonal development of non-linear functionals in series of Fourier-Hermite functionals. *Ann. Math.* 1947, **48**(2), 385–92.

26. Chung D. M., Ji U. C. & Saito K. Cauchy problems associated with the Lévy Laplacian in white noise analysis. *Infinite Dimensional Analysis, Quantum Probability and Related Topics.* 1999, **2**(1), 131–53.

27. Dalecky Yu. L. & Fomin S. V. *Measures and Differential Equations in Infinite-dimensional Space.* Amsterdam, New York: Kluwer Acad. Publ., 1991.

28. Dorfman I. Ya. On the mean values and the Laplacian of functions on Hilbert space. *Mat. Sbornik.* 1970, **81(123)**, 2, 192–208 (in Russian).

29. Dorfman I. Ya. On heat equation on Hilbert space. *Vestnik Moscow State Univ.* 1971, **4**, 46–51 (in Russian).

30. Dorfman I. Ya. On one class of vector-valued generalized functions and its applications to the theory of Lévy Laplacian. II. *Vestnik Moscow State Univ.* 1973, **5**, 18–25 (in Russian).

31. Dorfman I. Ya. Methods of theory of bundles in the theory of Lévy Laplacian. *Uspekhi Mat. Nauk.* 1973, **28**, 6, (174), 203–4 (in Russian).

32. Dorfman I. Ya. Methods of theory of bundles in the theory of Lévy Laplacian. *Izvestiya Akad. Nauk Armenian SSR. Mathematics.* 1974, **9**(6), 486–503 (in Russian).

33. Dorfman I. Ya. On divergence of vector fields on Hilbert space. *Sibirsk. Mat. Zhurn.* 1976, **17**(5), 1023–31 (in Russian).
34. Emch G.G. *Algebraic Methods in Statistical Mechanics and Quantum Field Theory.* New York, London, Sydney, Toronto: Wiley, 1972.
35. Etemadi N. On the laws of large numbers for nonnegative random variables. *J. Multivar. Anal.* 1983, **13**(1), 187–93.
36. Feller M. N. On the Laplace equation in the space $L_2(C)$. *Dokl. Akad. Nauk Ukrain. SSR.* 1965, **12**, 1558–62 (in Russian).
37. Feller M. N. On the Poisson equation in the space $L_2(C)$. *Dokl. Akad. Nauk Ukrain. SSR.* 1966, 4, 426–9 (in Russian).
38. Feller M. N. On the equation $\Delta U[x(t)] + P[x(t)]U[x(t)] = 0$ in a function space. *Dokl. Akad. Nauk SSSR.* 1967, **172**(6), 1282–5 (in Russian).
39. Feller M. N. On the equation $\Delta_{2m}U[x(t)] = 0$ in a function space. *Dokl. Akad. Nauk Ukrain. SSR.* 1967, Ser. A, 10, 879–83 (in Russian).
40. Feller M. N. On the polyharmonic equation in a function space. *Dokl. Akad. Nauk Ukrain. SSR.* 1968, Ser. A, 11, 1005–11 (in Russian).
41. Feller M. N. On one class of elliptic equations of higher orders in variational derivatives. *Dokl. Akad. Nauk Ukrain. SSR.* 1969, Ser. A. **12**, 1096–101 (in Russian).
42. Feller M. N. On one class of perturbed elliptic equations of higher orders in variational derivatives. *Dokl. Akad. Nauk Ukrain. SSR.* 1970, Ser. A. **12**, 1084–7 (in Russian).
43. Feller M. N. On infinite-dimensional elliptic operators. *Dokl. Akad. Nauk SSSR.* 1972, **205**(1), 36–9 (in Russian).
44. Feller M. N. On the solvability of infinite-dimensional elliptic equations with constant coefficients. *Dokl. Akad. Nauk SSSR.* 1974, **214**(1), 59–62 (in Russian).
45. Feller M. N. Infinite-dimensional elliptic equations of P. Lévy type. *Materials of All-Union School on Differential Equations with Infinite Number of Independent Variables.* Yerevan: Published by Akad. Nauk Armenian SSR. 1974, pp. 79–82 (in Russian).
46. Feller M. N. On the solvability of infinite-dimensional self-adjoint elliptic equations. *Dokl. Akad. Nauk SSSR.* 1975, **221**(5), 1046–9 (in Russian).
47. Feller M. N. A family of infinite-dimensional elliptic differential equations. *Proc. All-Union Conf. on Partial Differential Equations, on the occasion of the 75th birthday of academician I. G. Petrovski.* Moscow: Izdat. Moskov. Univ. 1978, pp. 467–8 (in Russian).
48. Feller M. N. On the solvability of infinite-dimensional elliptic equations with variable coefficients. *Mat. Zametki.* 1979, **25**(3), 419–24 (in Russian).
49. Feller M. N. Family of infinite-dimensional elliptic expressions. *Spectral Theory of Operators, Proceedings of All-Union Mathematical School.* Baku: Elm, 1979, pp. 175–82 (in Russian).
50. Feller M. N. Infinite-dimensional Laplace–Lévy differential operators. *Ukrain. Mat. Zhurn.* 1980, **32**(1), 69–79 (in Russian).
51. Feller M. N. Infinite-dimensional self-adjoint Laplace–Lévy differential operators. *Ukrain. Mat. Zhurn.* 1983, **35**(2), 200–6 (in Russian).
52. Feller M. N. Infinite-dimensional elliptic equations and operators of Lévy type. *Russian Math. Surveys.* 1986, **41**(4), 119–70.

53. Feller M. N. Self-conjugacy of nonsymmetrized infinite-dimensional operator of Laplace–Lévy. *Ukrain. Mat. Zhurn.* 1989, **41**(7), 997–1001 (in Russian).
54. Feller M. N. Harmonic functions of infinite number of variables. *Uspekhi Mat. Nauk.* 1990, **45**(4), 112–3 (in Russian).
55. Feller M. N. Reserve of harmonic functions of the infinite number of variables. I. *Ukrain. Mat. Zhurn.* 1990, **42**(11), 1576–9 (in Russian).
56. Feller M. N. Reserve of harmonic functions of the infinite number of variables. II. *Ukrain. Mat. Zhurn.* 1990, **42**(12), 1687–93 (in Russian).
57. Feller M. N. Reserve of harmonic functions of the infinite number of variables. III. *Ukrain. Mat. Zhurn.* 1992, **44**(3), 417–23 (in Russian).
58. Feller M. N. Necessary and sufficient conditions for a function of infinitely many variables to be harmonic (the Jacobi case). *Ukrain. Mat. Zhurn.* 1994, **46**(6), 785–8 (in Russian).
59. Feller M. N. One more condition of harmonicity of functions of infinite number of variables (translationally nonpositive case). *Ukrain. Mat. Zhurn.* 1994, **46**(11), 1602–5 (in Russian).
60. Feller M. N. On one nonlinear equation not solved with respect to Lévy Laplacian. *Ukrain. Mat. Zhurn.* 1996, **48**(5), 719–21 (in Russian).
61. Feller M. N. The Riquier problem for a nonlinear equation solved with respect to the iterated Lévy Laplacian. *Ukrain. Mat. Zhurn.* 1998, **50**(11), 1574–7 (in Russian).
62. Feller M. N. The Riquier problem for a nonlinear equation unresolved with respect to the iterated Lévy Laplacian. *Ukrain. Mat. Zhurn.* 1999, **51**(3), 423–7 (in Russian).
63. Feller M. N. Notes on infinite-dimensional non-linear elliptic equations. I. *Spectral and Evolutionary Problems. Proceedings of the Eight Crimean Autumn Math. School–Symposium.* 1998, Vol. 8, pp. 182–187.
64. Feller M. N. Notes on infinite-dimensional non-linear elliptic equations. II. *Spectral and Evolutionary Problems. Proceedings of the Ninth Crimean Autumn Math. School–Symposium.* 1999, Vol. 9, pp. 177–181.
65. Feller M. N. Cauchy problem for non-linear equations involving the Lévy Laplacian with separable variables. *Methods of Functional Analysis and Topology.* 1999, **5**(4), 9–14.
66. Feller M. N. Notes on infinite-dimensional nonlinear parabolic equations. *Ukrain. Mat. Zhurn.* 2000, **52**(5), 690–701 (in Russian).
67. Fukusima M., Oshima Y. & Takeda M. *Dirichlet Forms and Symmetric Markov Processes.* Berlin, New York: WG, 1994.
68. Gâteaux R. Sur la notion d'intégrale dans le domaine fonctionnel et sur la théorie du potentiel. *Bull. Soc. Math. France.* 1919, **47**, 47–70.
69. Gromov M. & Milman V. D. A topological application of the isoperimetric inequality. *Am. J. Math.* 1983, **105**(4), 843–54.
70. Gross L. Potential theory on Hilbert space. *J. Funct. Anal.* 1967, **1**(2), 123–81.
71. Haag R. The mathematical structure of the Bardeen–Cooper–Schriffer Model. *Nuovo cimento.* 1962, **25**(2), 287–99.
72. Hasegawa Y. Lévy's functional analysis in terms of an infinite dimensional Brownian motion. I. *Osaka J. Math.* 1982, **19**(2), 405–28.

73. Hasegawa Y. Lévy's functional analysis in terms of an infinite dimensional Brownian motion. II. *Osaka J. Math.* 1982, **19**(3), 549–70.
74. Hasegawa Y. Lévy's functional analysis in terms of an infinite dimensional Brownian motion. III. *Nagoya Math. J.* 1983, **90**, 155–73.
75. Hida T. Brownian Motion. New York, Heidelberg, Berlin: Springer, 1980.
76. Hida T. Brownian motion and its functionals. *Ricerche di Matematica*. 1985, **34**, 183–222.
77. Hida T. A role of the Lévy Laplacian in the causal calculus of generalized white noise functionals. In: *Stochastic Processes*. Springer-Verlag. 1993, pp. 131–139.
78. Hida T. Random fields as generalized white noise functionals. *Acta Applicandue Math*. 1994, **35**, 49–61.
79. Hida T., Saito K. White noise analysis and the Lévy Laplacian. In: *Stochastic Processes in Physics and Engineering*. Dordrecht, Boston, Lancaster, Tokyo: D. Reidel Pub. Co., 1988, pp. 177–84.
80. Hida T., Kuo H.-H., Potthoff J. & Streit L. *White Noise. An Infinite Dimensional Calculus*. Dordrecht, Boston, London: Kluwer Acad. Pub., 1993.
81. Ito K. Multiple Wiener integral. *J. Math. Soc. Japan* 1951, **3**(1), 157–69.
82. Kalinin V. V. Laplace and Poisson equations in functional derivatives in Hilbert space. *Izvestiya Vuzov. Mathematics*. 1972, 3, 20–2 (in Russian).
83. Koshkin S. V. Lévy-like continual means on space L_p. *Methods of Functional Analysis and Topology* 1998, **4**(2), 53–65.
84. Koshkin S. V. Asymptotic concentration of Gibbs–Vlasov measures. *Methods of Functional Analysis and Topology* 1998, **4**(4), 40–9.
85. Kubo I. & Takenaka S. Calculus on Gaussian white noise. IV. *Proc. Japan. Acad.* 1982, **58**, Ser. A, 186–9.
86. Kuo H.-H. On Laplacian operators of generalized Brownian Functionals. In: *Lecture Notes in Math*. Berlin: Springer-Verlag, 1986, Vol. 1203, pp. 119–128.
87. Kuo H.-H. Brownian motion, diffusions and infinite dimensional calculus. In: *Lecture Notes in Math*. Berlin: Springer-Verlag, 1988, Vol. 1316, pp. 130–169.
88. Kuo H.-H. *White Noise Distribution Theory*. Boca Raton, New York, London: CRC Press, 1996.
89. Kuo H.-H, Obata N. & Saito K. Lévy Laplacian of generalized functions on a nuclear space. *J. Func. Anal.* 1990, **94**(1), 74–92.
90. Kuo H.-H, Obata N. & Saito K. Diagonalization of the Lévy Laplacian and related stable processes. *Infinite Dimensional Analysis, Quantum Probability and Related Topics* 2002. **5**(3), 317–31.
91. Léandre R. & Volovich I. A. The stochastic Lévy Laplacian and Yang–Mills equation on manifolds. *Infinite Dimensional Analysis, Quantum Probability and Related Topics* 2001, **4**(2), 161–72.
92. Lévy P. Sur la généralisation de l'équation de Laplace dans le domaine foctionnel. *C.R. Acad. Sc.* 1919, **168**, 752–55.
93. Lévy P. *Leçons d'analyse fonctionnelle*. Paris: Gauthier–Villars, 1922.
94. Lévy P. Analyse fonctionnelle. *Mem. Sc. Math. Acad. Sc. Paris* 1925, Fasc. V, 1–56.
95. Lévy P. *Problémes concrets d'analyse fonctionnelle*. Paris: Gauthier–Villars, 1951.

96. Lévy P. Random functions: a Laplacian random function depending on a point of Hilbert space. *Univ. Calif. Publ. Statistics* 1956, **2**(10), 195–205.
97. Milman V. D. Diameter of a minimal invariant subset of equivariant Lipschitz Actions on Compact Subsets of R^k. In: *Lecture Notes in Math*. Berlin: Springer-Verlag, 1987, Vol. 1267, pp. 13–20.
98. Milman V. D. The heritage of P.Lévy in geometrical functional analysis. *Asterisque* 1988, 157–158, 273–301.
99. Naroditsky V. A. On operators of Laplace-Lévy type. *Ukrain. Mat. Zhurn.* 1977. **29**(5), 667–73 (in Russian).
100. Nemirovsky A. S. To general theory of Laplace-Lévy operator. *Funktsional'nyi Analiz i ego Prilozheniya*. 1974, **8**(3), 79–80 (in Russian).
101. Nemirovsky A. S. Laplace-Lévy operator on one class of functions. *Trudy Mosk. Mat. Obshchestva*. 1975, **32**, 227–50 (in Russian).
102. Nemirovsky A. S. & Shilov G. E. On axiomatic description of Laplace operator for functions on Hilbert space. *Funktsional'nyi Analiz i ego Prilozheniya* 1969, **3**(3), 79–85 (in Russian).
103. Obata N. A note on certain permutation groups in the infinite dimensional rotation group. *Nagoya Math. J.* 1988, **109**, 91–107.
104. Obata N. The Lévy Laplacian and mean value theorem. In: *Lecture Notes in Math*. Berlin: Springer-Verlag, 1989, Vol. 1379, pp. 242–253.
105. Obata N. A characterization of the Lévy Laplacian in terms of infinite dimensional rotation groups. *Nagoya Math. J.* 1990, **118**, 111–32.
106. Obata N. Quadratic quantum white noises and Lévy Laplacian. *Nonlinear Analysis*. 2001, **47**, 2437–48.
107. Owchar M. Wiener integrals of multiple variations. *Proc. Am. Math. Soc.* 1952, **3**, 459–70.
108. Petrov V. V. On order of growth of sums of dependent random variables. *Teoriya Veroyatnostei i ee Primeneniya*. 1973, **18**(2), 358–60 (in Russian).
109. Petrov V. V. On strong law of large numbers for sequence of orthogonal random values. *Vestnik Leningradsk. Univ.* 1975, 7, 52–7 (in Russian).
110. Petrov V. V. *Limit Theorems for Sums of Independent Random Variables*. Moscow: Nauka, 1987 (in Russian).
111. Polishchuk E. M. Continual means and harmonic functionals. *Uspekhi Mat. Nauk*. 1960, **15**(3) (93), 229–31 (in Russian).
112. Polishchuk E. M. On continual means and singular distributions. *Teoriya Veroyatnostei i ee Primeneniya* 1961, **6**(4), 465–9 (in Russian).
113. Polishchuk E. M. On expansion of continual means in terms of degrees of functional Laplacian. *Sibirsk. Mat. Zhurn*. 1962, **3**(6), 852–69 (in Russian).
114. Polishchuk E. M. Functionals which are orthogonal on sphere. *Sibirsk. Mat. Zhurn*. 1963, **4**(1), 187–205 (in Russian).
115. Polishchuk E. M. On differential equations with functional parameters. *Ukrain. Mat. Zhurn*. 1963, **15**(1), 13–24 (in Russian).
116. Polishchuk E. M. On functional Laplacian and equations of parabolic type. *Uspekhi Mat. Nauk*. 1964, **19**(2) (116), 155–62 (in Russian).
117. Polishchuk E. M. Linear equations in functional Laplacians. *Uspekhi Mat. Nauk*. 1964, **19**(2) (116), 163–70 (in Russian).

118. Polishchuk E. M. On functional analogues of heat equations. *Sibirsk. Mat. Zhurn.* 1965, **6**(6), 1322–31 (in Russian).

119. Polishchuk E. M. On equations of Laplace and Poisson type in functional space. *Mat. Sbornik.* 1967, **72(114)** 2, 261–92 (in Russian).

120. Polishuk E. M. Quelques théorèmes sur les valeurs moyennes dans les domaines fonctionnels. *Bull. Sci. Math. 2 série.* 1969, **93**, 145–56.

121. Polishchuk E. M. On some new interrelations of controlled systems with equations of mathematical physics. I. *Differentsial'nye Uravneniya* 1972, **8**(2), 333–48 (in Russian).

122. Polishchuk E. M. On some new interrelations of controlled systems with equations of mathematical physics. II. *Differentsial'nye Uravneniya* 1972, **8**(5), 857–70 (in Russian).

123. Polishchuk E. M. Elliptic operators on rings of functionals. *Funktsional'nyi Analiz i ego Prilozheniya* 1974, **8**(1), 31–5 (in Russian).

124. Polishchuk E. M. Operator $\sum_1^\infty (\partial^2/\partial c_k \partial \bar{c}_k)$ in space with kernel function. *Litovskii Mat. Sbornik* (Lithuanian Mat. Collection) 1976, **16**(3), 123–36 (in Russian).

125. Polishchuk E. M. *Continual Means and Boundary Value Problems in Function Spaces.* Berlin: Akademie-Verlag. 1988.

126. Saito K. Ito's formula and Lévy's Laplacian. *Nagoya Math. J.* 1987, **108**, 67–76.

127. Saito K. Ito's formula and Lévy's Laplacian. II. *Nagoya Math. J.* 1991, **123**, 153–69.

128. Saito K. A C_0 − group generated by the Lévy Laplacian. *J. Stoch. An. Appl.* 1998, **16**, 567–84.

129. Saito K. A C_0 − group generated by the Lévy Laplacian. II. *Infinite Dimensional Analysis, Quantum Probability and Related Topics* 1998, **1**(3), 425–37.

130. Saito K. & Tsoi A. H. The Lévy Laplacian acting on Poisson noise functionals. *Infinite Dimensional Analysis, Quantum Probability and Related Topics* 1999, **2**(4), 503–10.

131. Scarlatti S. On some new type of infinite dimensional Laplacians. *Progress in Probability* 1999, **45**, 267–74.

132. Shilov G. E. On some problems of analysis in Hilbert space. I. *Funktsional'nyi Analiz i ego Prilozheniya* 1967, **1**(2), 81–90 (in Russian).

133. Shilov G. E. On some problems of analysis in Hilbert space. II. *Mat. Issledovaniya* Published by Acad. Sci. of Moldavian SSR. 1968, **2**(4), 166–86 (in Russian).

134. Shilov G. E. On some problems of analysis in Hilbert space. III. *Mat. Sbornik.* 1967, **74(116)**, (1), 161–8 (in Russian).

135. Shilov G. E. On some solved and unsolved problems of theory of functions on Hilbert space. *Vestnik Moscow State Univ.* 1970, **2**, 66–8 (in Russian).

136. Shilov G. E. & Phan Dich Tinh. *Integral, Measure, and Derivative on Linear Spaces.* Moscow: Nauka, 1967 (in Russian).

137. Sikiryavyi V. Ya. On solving of boundary-value problems associated with operators of pseudospherical differentiation. *Uspekhi Mat. Nauk.* 1970, **25**(6) (156), 231–2 (in Russian).

138. Sikiryavyi V. Ya. To theory of boundary-value problems for equations with operator of pseudodifferentiation. *Uspekhi Mat. Nauk.* 1971, **26**(4) (160), 247–8 (in Russian).

139. Sikiryavyi V. Ya. Constructive description of axiomatic Laplace operator for functions on Hilbert space. *Vestnik Moscow State Univ.* 1971, **2**, 83–6 (in Russian).
140. Sikiryavyi V. Ya. Operator of quasidifferentiation and boundary-value problems related to it. *Trudy Mosk. Mat. Obshchestva* 1972, **27**, 195–246 (in Russian).
141. Sikiryavyi V. Ya. Invariant Laplace operator as operator of pseudospherical differentiation. *Vestnik Moscow State Univ.* 1972, **3**, 66–73 (in Russian).
142. Sikiryavyi V. Ya. On boundary-value problems for multiplicatively generated operator of quasidifferentiation. *Ukrain. Mat. Zhurn.* 1973, **25**(3), 406–9 (in Russian).
143. Sikiryavyi V. Ya. Boundary-value problems for multiplicatively generated operator of quasidifferentiation. *Sibirsk. Mat. Zhurn.* 1973, **14**(6), 1313–20 (in Russian).
144. Sikiryavyi V. Ya. On sufficient conditions of existence of means of Gâteaux–Lévy. *Uspekhi Mat. Nauk.* 1974, **29**(5) (179), 239–40 (in Russian).
145. Sikiryavyi V. Ya. Linear quasidifferential equations. *Ukrain. Mat. Zhurn.* 1975, **27**(1), 121–7 (in Russian).
146. Sokolovsky V. B. Second and third boundary-value problems in Hilbert ball for equations of elliptic type solved relative to functional Laplacian. *Izvestiya Vuzov. Mathematics* 1975, **3**, 111–4 (in Russian).
147. Sokolovsky V. B. Boundary-value problem without initial conditions for one equation with functional Laplacian in Hilbert space. *Izvestiya Vuzov. Mathematics* 1975, **11**, 102–5 (in Russian).
148. Sokolovsky V. B. Second boundary-value problem without initial conditions for heat equation in Hilbert ball. *Izvestiya Vuzov. Mathematics* 1976, **5**, 119–23 (in Russian).
149. Sokolovsky V. B. New relations of some problems for equations with Laplace–Lévy operators with problems of mathematical physics. *Izvestiya Vuzov. Mathematics* 1980, **11**, 82–4 (in Russian).
150. Sokolovsky V. B. Infinite-dimensional equations with Lévy Laplacian and some variational problems. *Ukrain. Mat. Zhurn.* 1990, **42**(3), 398–401 (in Russian).
151. Sokolovsky V. B. Infinite-dimensional parabolic equations with Lévy Laplacian and some variational problems. *Sibirsk. Mat. Zhurn.* 1994, **35**(1), 177–80 (in Russian).
152. Thirring W. & Wehrl A. On the mathematical structure of the B.C.S.-Model. *Commun. Math. Phys.* 1967, **4**(5), 303–14.
153. Vershik A. M. The general theory of Gaussian measures in linear spaces. *Uspekhi Mat. Nauk.* 1964, **19**(1), 210–2 (in Russian).
154. Volterra V. Sulle equazioni alle derivate fonzionall. Atti della Reale Accademia dei Lincei, 1914, 5 serie, **23**, 393–9.
155. Wehrl A. Spin waves and the BCS-Model. *Commun. Math. Phys.* 1971, **23**(4), 319–42.
156. Wiener N. Differencial space. *J. Math. Phys.* 1923, **2**(3), 131–74.
157. Wiener N. *Nonlinear Problems in Random Theory*. New York: Technology Press of MIT and Wiley, 1958.
158. Yadrenko M. I. *Spectral Theory of Random Fields*. Kiev: Vyshcha Shkola, 1980 (in Russian).
159. Zhang Y. Lévy Laplacian and Brownian particles in Hilbert spaces. *J. Func. Anal.* 1995, **133**(2), 425–41.

Index